T4-ATW-902

# *Experiments That Explore*

## *Recycling*

# EXPERIMENTS THAT EXPLORE RECYCLING

*An Investigate! Book*
*by Martin J. Gutnik*

*Illustrated by Sharon Lane Holm*

*The Millbrook Press*
*Brookfield, Connecticut*

Cover photograph courtesy of Superstock

Photographs courtesy of: AP/Wide World: frontis, p. 59; Tennessee Valley Authority: p. 8; Environmental Action Fund: p. 15; Photo Researchers: pp. 29 (Alvin E. Steffan, National Audubon Society), 44 (Lynwood M. Chace, National Audubon Society).

Library of Congress Cataloging-in-Publication Data
Gutnik, Martin J.
Experiments that explore recycling / by Martin J. Gutnik. —
Millbrook Press Library ed.
p. cm. — (An Investigate! book)
Includes bibliographical references and index.
Summary: Uses experiments to demonstrate the effects of dumping solid waste into our environment and explores what can be done about it.
ISBN 1-56294-116-X
1. Refuse and refuse disposal—Juvenile literature. 2. Recycling (Waste, etc.)—Juvenile literature. 3. Ecology—Juvenile literature. 4. Refuse and refuse disposal—Experiments—Juvenile literature. 5. Recycling (Waste, etc.)—Experiments—Juvenile literature. 6. Ecology—Experiments—Juvenile literature.
[1. Refuse and refuse disposal—Experiments. 2. Recycling (Waste)—Experiments. 3. Experiments.] I. Title. II. Series.
TD792.G87 1992
628.4′458—dc20 91-26147 CIP AC


Printed in the United States of America
5 4 3 2 1

# *Contents*

*To my grandson, Justin Michael Weimer.*
*May he grow up in a clean, safe environment.*

*Thanks to my wife Natalie for doing the*
*primary research for this book.*

# *Experiments That Explore Recycling*

*For many people, large dams have become symbolic of a misguided attitude that we can control and use nature in any way we wish, without serious consequences.*

# 1

## *Introduction*

Throughout history, people have demonstrated a surprising lack of understanding concerning their environment. They thought that nature would always provide for them. Even after learning that the planet's resources are limited, many still continued to have this same attitude, believing that technology would allow them to "have it all." Dams could be built to control rivers, guns and poisons would get rid of dangerous predators and insect pests, lakes could be dug anywhere, nuclear energy would fulfill all energy needs, and landfills would absorb all the garbage. People felt confident that they would always be in control of their environment. Eventually, people began to regard themselves as no longer part of the natural order, as separate from the ecological environment and not dependent on the balance of nature.

In fact, just the opposite is true. Today, more than ever, we rely upon nature to fix things for us. Because of our self-centered thinking, we have created an environmental crisis that now threatens our very existence on this planet. Our technologies, while making life more comfortable for us, have allowed us to create billions of tons of material that must somehow be disposed of.

A huge number of new products for human consumption and convenience are introduced each year. They bring with them new containers, new chemicals, and more waste. As a consequence, we have a garbage crisis. The garbage dilemma has become a major environmental problem. A top priority today is deciding what to do with all of our waste.

We are a "throwaway" society. Every day, the people of the industrialized world throw out everything from toothpicks to toothpaste tubes to old television sets. On the average, every person throws out approximately 4.5 pounds of solid waste per day. If you multiply this by 365 days per year, and then by 250 million people, you can see that in the United States alone we throw out approximately 410 billion pounds of solid waste per year. This is enough waste to fill a line of garbage trucks reaching one fifth of the way from the Earth to the moon.

Most of this waste ends up in *sanitary landfills,* places where solid waste is dumped, compacted into layers, and covered with soil. The products that we throw away are using up many of our natural resources, and we also have problems with landfills. Many are leaking toxic, or poisonous, fumes into the air and toxic chemicals into underground water systems. The poisons from landfills contaminate our water and make it unfit for human and animal consumption. Since all the water on Earth flows and interacts with all other water systems, these poisons eventually spread to other water resources and contaminate them.

This book will explore the waste problem by demonstrating, through science projects, the effects of dumping solid waste into our environment. After analyzing the problem, the book will explore what can be done, stressing the three "Rs":

*reduce* the amount of waste produced;
*reuse* items; and
*recycle* waste materials.

## THE SCIENTIFIC METHOD

A science project can explore a problem or an idea in any scientific discipline. Environmental projects involve the fields of biology (life science), physics (physical science), and chemistry (chemical science).

*Ecology* is the study of how all living things interrelate with each other and their nonliving environment. Because environmental science spans all fields of science, ecologists (people who study ecology) must have broad scientific knowledge in all the scientific disciplines. They must be familiar with the important interrelationships that occur in nature.

Science projects, to be meaningful to the researcher and to others, must proceed in a logical order. People involved in science projects usually follow the scientific method of discovery. This is a series of steps that most trained experimenters use for investigating a problem.

Science projects begin with observing an event or a phenomenon. *Observation* means using all your senses to find out all you can. An investigator in a science project must limit observations to the subject he or she is investigating. Since many things cannot be observed by our senses alone, scientists and other researchers use special instruments to help them observe. A microscope helps a medical researcher study the makeup of cells. Telescopes help astronomers study the planets and stars. Barometers, thermometers, and radar help meteorologists study weather patterns.

Previous knowledge that has been gathered by others, from observations and experiments, is passed on in books, magazines, and journals. Researching this scientific literature is considered part of the observation (research) process.

After completing the observation phase, the researcher must organize the information gathered. That is, the information must be classified. *Classification* means grouping similar objects and related pieces of information. This gives us important new information.

Relevant observations and classifications concerning various ecological phenomena have been included as part of the background material that precedes each of the experiments in this book. Reading this background material carefully will enable you to understand the additional steps in the scientific process as they are formally presented in the experiments.

Once events have been observed and classified, the researcher must pose a question or a problem that is directly related to what has been observed. This is called either an *inference* or a *prediction*. An inference is a statement about something that has happened. A prediction is a statement about something that is likely to happen.

The inference or prediction leads the researcher to make an educated guess concerning possible answers to the problem. This guess is offered as a *hypothesis*. A hypothesis is an inference or a prediction that can be tested. It is usually phrased as an if-then statement (for example: *If* a whale is a mammal, *then* it must have hair on its body). The hypothesis gives direction to the scientific investigation. Everything done in an experiment or a research project must relate to proving or disproving the hypothesis.

The *results* of an investigation should be recorded. This will provide a ready, documented source upon which to base any *conclusions*. The conclusion states whether the hypothesis was proved or disproved. It must also tell why the hypothesis was correct or incorrect.

Often, the hypothesis will not be valid because of *variables* that have not been controlled. Variables are differing conditions that may alter the results of your experiment or research. It is important to state the variables in your conclusion, so that you and other researchers may better analyze the results of your investigation.

Many times the results of an experiment will not prove or disprove your hypothesis. This does not mean that the experiment has failed. Rather, it usually indicates a need for further investigation. When investigating the problem again, you should restate the hypothesis and redesign the experiment.

# 2

## *Understanding Ecosystems*

The *biosphere* is any place on Earth where life can exist. There are only four nonliving parts of the biosphere. They are air, water, soil, and energy (especially light energy). When all four parts of the biosphere are found together in a proper mix, life can exist.

Around the world, there are large areas of similar climate, vegetation, and wildlife. These areas are called *biomes*. Some examples of biomes are rain forests, deserts, grasslands, deciduous forests, and coniferous forests.

Within each biome are smaller, more specialized, areas called *ecosystems*. Some examples of ecosystems are a pond, a woodlot, a bog, and a marsh. Ecosystems can vary as much as do species of plants or animals.

All biomes and ecosystems are interrelated by sharing the same air, water, and soil. The various forces of nature, such as wind, precipitation (rain, snow, sleet, or hail), erosion, and water movement, interconnect all of these systems. They make Earth an interdependent, living planet.

Because of this interdependence, the Earth is sensitive to the activities of nature and people. When a natural event causes pollution, it is called *natural pollution*. Natural pollution is not as devastating to the environment as pollution caused by people because it is usually short-lived and often localized. Volcanoes emitting (releasing) sulfur dioxide gas into the air, bogs giving off hydrogen sulfide gas, and pollen distribution are all examples of natural pollution. The harmful effects of these kinds of pollution are neutralized by nature itself. Because they do not occur constantly or affect wide areas, their impact on the global environment is usually small.

Human-made pollution is often devastating to the natural environment. The activities of people, because they are widespread and ongoing, interrupt and alter the natural cycles on Earth. Preventing further pollution of the environment is the most challenging problem facing the human race today.

Human garbage is a form of pollution stemming from the activities of individuals, communities, and nations. According to some estimates, in 1987 Americans created enough garbage to cover, 6 feet deep, a four-lane highway stretching from New York to Los Angeles. Because of the massive amount of waste, we are rapidly running out of landfill space and other areas in which to dispose of it.

*Incineration* (burning) of the garbage is probably not an answer. It creates toxic gases and ash, which pollute the air.

Perhaps the best way to handle this problem is to create less waste. If each individual took the responsibility to reduce the amount of garbage he or she produces, we, as a nation, could cut the amount of garbage produced by at least one third.

## THE CARBON CYCLE

Living organisms need many different chemical elements in order to stay alive and grow. Some of these elements are carbon, oxygen, phosphorus, and potassium. These materials are passed back and forth between the living and nonliving aspects of the environment. Their movement is almost circular. Therefore, their patterns of movement are termed *cycles*. Nature's cycles differ from one another in how fast elements are exchanged and recycled. All of the cycles involve water plus gases or soils.

*City dumps are breeding grounds for rats, disease-carrying insects, and other pests.*

One element required by all living things is carbon (C). Carbon must find its way to all the living organisms in any ecosystem. The source of nearly all of the carbon in an ecosystem is carbon dioxide gas ($CO_2$), found free in the air and dissolved in water. Carbon dioxide gas must be transferred to all living things. This is accomplished through the *food web*. The food web is the way that energy, in the form of food, is passed around an ecosystem. The *carbon cycle* works through the food web.

The most basic activity in the food web, and thus in the carbon cycle, is *photosynthesis*. Photosynthesis is the process by which green plants make food (glucose, a simple sugar) from water, carbon dioxide gas, and light energy. Green plants change this simple sugar into various fats and starches, which they use for energy and nutrition. The food energy and stored carbon in plants are transferred to herbivores (animals that only eat green plants), then to carnivores (meat eaters) and omnivores (animals that eat both meat and green plants). The animals use some of the carbon for their own life processes and return most of the rest to the air or water through respiration (breathing). People breathe in oxygen and breathe out carbon dioxide gas.

Some of the carbon left in the plant tissues is eventually released through the process of decay. Plant and animal waste and dead material are *decomposed,* or broken down, into simpler substances. Most of the carbon that was locked up in these living things is then returned to the environment as carbon dioxide gas.

## Science Project #1—
## Nature's Interrelationships:
## The Carbon Cycle

### *Materials Needed*

- 2 coleus, impatiens, or geranium plants (can be obtained from a garden center)
- 2 large bell, pickle, or mayonnaise jars
- a few tablespoons of soda lime crystals (a carbon dioxide absorbent)
- 4″ tall glass jar
- pitcher of water

(NOTE: The scientific equipment and materials used in this and the other science experiments in this book can usually be obtained from your school science department.)

***Inference.*** All living things need carbon in order to exist.

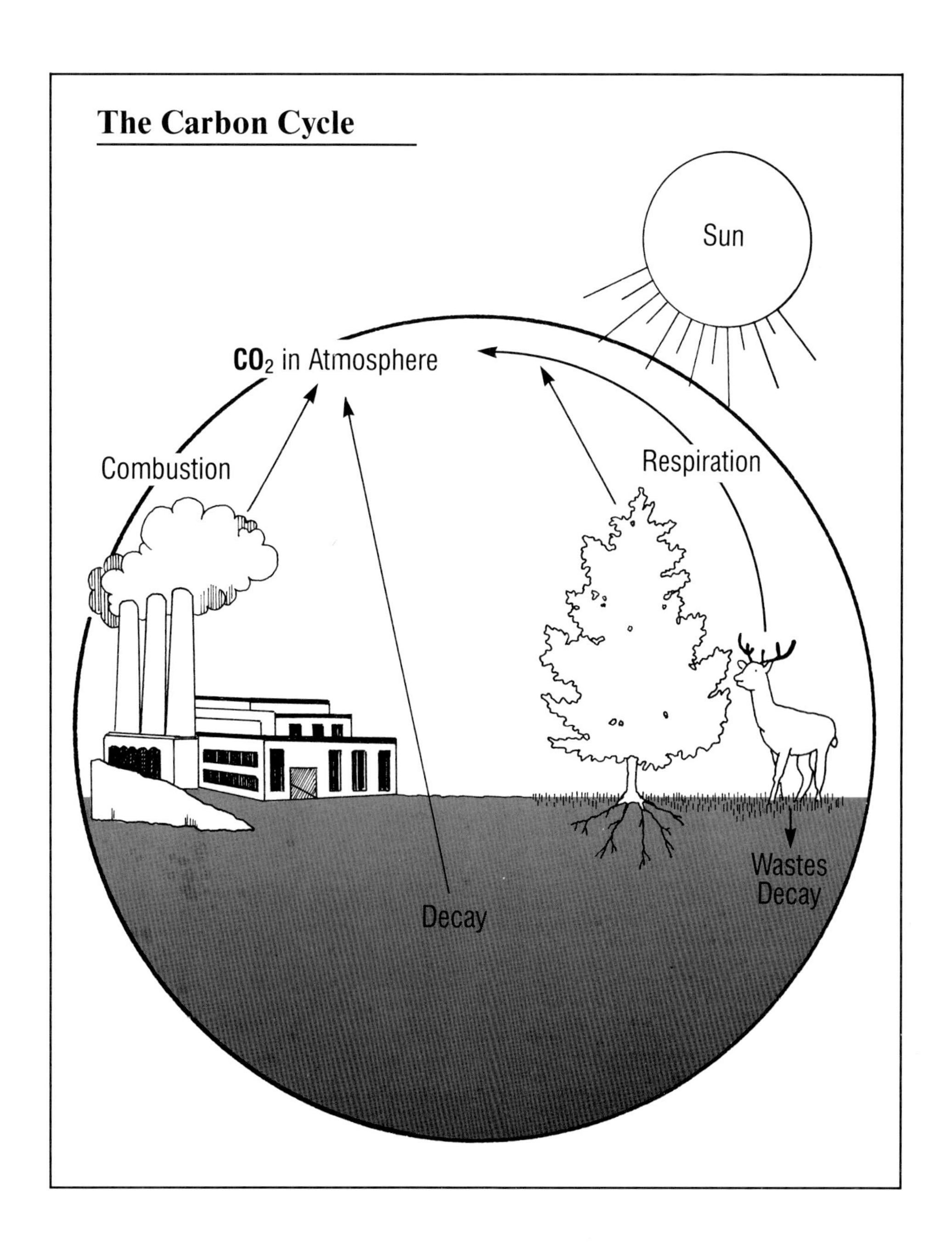
The Carbon Cycle
Sun
$CO_2$ in Atmosphere
Combustion
Respiration
Decay
Wastes
Decay

## Demonstrating the Carbon Cycle

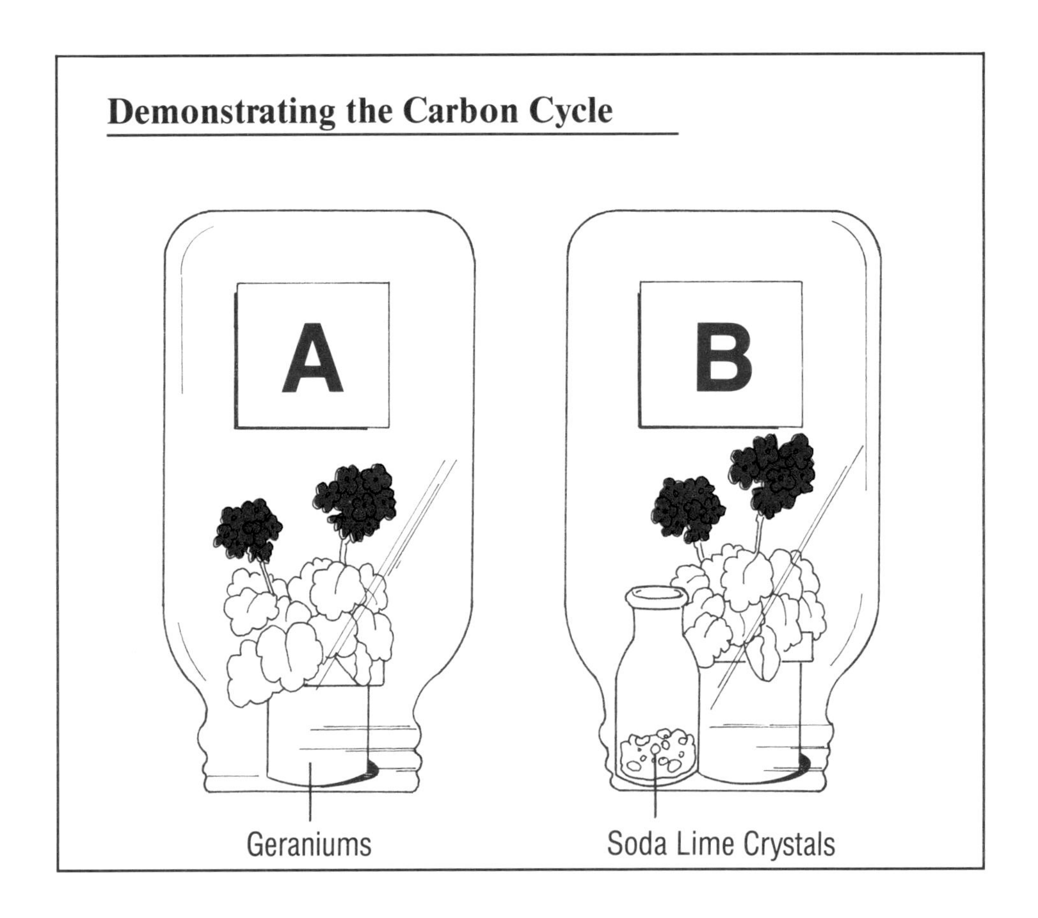


***Hypothesis.*** If living things are denied carbon, then they will eventually die.

***Procedure.*** Water two coleus, impatiens, or geranium plants. Put each plant where it will receive plenty of light. Cover one plant with a bell jar (or large pickle or mayonnaise jar). Label this jar "*A*." Pour soda lime crystals into the 4-inch jar and place this jar next to the uncovered plant. Cover both the plant and the soda lime with the other bell jar. Label this jar "*B*."

Observe the plants for one month. Once each week record how each plant is doing. Change the soda lime after two weeks. The soda lime should absorb the carbon dioxide available to the plant and prevent it from carrying on photosynthesis.

***Conclusion.*** My hypothesis was correct because plant *B* could not carry on the process of photosynthesis and thus died.

(NOTE: Conclusions to the rest of the projects will be presented in a special section at the end of this book.)

**Chart of Results**

| | Plant "A" | Plant "B" |
|---|---|---|
| Week 1 | Doing well, green, healthy | Doing well, green, healthy |
| Week 2 | | |

**(MAKE YOUR OWN CHART OR PHOTOCOPY THIS ONE.)**

# 3

# *Defining Solid Waste*

The garbage crisis is not a new problem. Throughout history, ever since people started to live in communities, societies have had a garbage problem. As the human population has grown, the garbage problem has increased. Today, with 5.6 billion people on Earth and a throwaway attitude in modern, wealthy societies, we have a monumental problem. How do we dispose of our trash without burying ourselves in it?

*Solid waste* is a term that refers to all kinds of solid and semisolid wastes, including household garbage, yard wastes, ashes, industrial waste, swill (slop or liquid garbage), demolition and construction wastes, and household discards such as appliances and furniture.

By some estimates, in the United States we throw away enough paper each year to build a 12-foot-high wall from Los Angeles to New York City and enough garbage each day to fill 63,000 garbage trucks that hold 7 to 14 tons of trash each.

In New York City alone, there are 7.2 million people who together generate 36,000 tons of solid waste each day. This waste comes from homes, schools, businesses, industries, government offices and facilities, stores, and restaurants. Most of the waste goes to a landfill on Staten Island. This landfill currently contains 36 billion square feet of trash.

Each man, woman, and child in the United States uses the equivalent of 600 pounds of paper each year (approximately five trees' worth).

In the Soviet Union, each person uses 25 pounds, and in China only 2 pounds are used per person.

If you were to analyze the solid waste produced by the state of Wisconsin, you would find the following: *Paper*—the equivalent of 25 million trees per year is thrown out; *metal cans*—the residents of the state throw away 2 billion cans per year, over half of which have been used for soda and beer; *glass containers*—over a billion per year are disposed of; *plastic bottles*—300 million per year; *tires*—4 million per year; *yard and food wastes*—500,000 tons a year; *other wastes, including pulp and paper mill sludges*—1.3 million tons a year; *ashes*—from electricity generation, 900,000 tons a year; and *foundry sands*—600,000 tons a year.

Wisconsin's trash is typical of the solid-waste problem in most states. Although some of the particulars may vary, the amounts for other states are just as staggering.

Solid waste is made up of many different materials. The most common items are:

*Glass*. Clear glass and colored glass are used for soda bottles, beer bottles, juice bottles, food jars, and light bulbs. Each year, approximately 45 billion glass bottles and jars are put into landfills in the United States.

*Paper*. This includes newspaper, office and notebook paper, and corrugated (ribbed) cardboard. Each year in the United States, approximately 3.75 billion tons of paper are disposed of. This is the equivalent of 1 billion trees annually.

*Metals*. Over 30 billion steel cans and 750 billion aluminum cans are trashed annually.

*Plastics*. There are many different kinds of plastics. In the United States, people use approximately 20 billion plastic containers each year. Plastics degrade (break down) much more slowly than paper and other items in landfills. They also seem to find their way more frequently into aquatic (water) environments, where they cause injury and death to aquatic animals. Even the so-called *biodegradable* plastic does not break down in landfills because there is no sunlight and air to help decompose it.

*Tires.* People living in the United States throw out approximately 200 million tires annually.

*Food and yard waste.* The average individual in the United States tosses out 210 pounds of food and yard wastes each year. Nationally, this amounts to 50 billion pounds of this waste annually.

## Science Project #2— Making People Aware of How Much Solid Waste They Generate

### *Materials Needed*

- 4.5 pounds of household trash, excluding food products (rinse thoroughly all the bottles and cans)
- large plastic garbage bag
- bathroom scale
- newspapers
- calculator
- trash from classroom
- trash from school lunchroom

***Prediction.*** When people better understand the solid-waste problem, they will be better prepared to make a commitment to *recycling,* or making new products from old ones, and to cutting down on waste.

***Hypothesis.*** If people are made aware of all the solid waste they generate, then they will better understand the need for waste reduction and recycling.

***Procedure.*** Fill a plastic bag with about 4.5 pounds of trash from your home. Include in this trash clean empty cans, bottles, paper, etc. First weigh yourself. Now pick up the trash bag and stand on the scale again. The weight should total your weight plus 4.5 pounds. If it is more or less, adjust the amount of trash in the bag until the scale shows the correct weight.

Bring the bag of trash to school and show it to your classmates. Explain that this is the average amount of trash produced daily by each American. Allow some of your classmates to lift the bag so that they can see how heavy it is. Spread some newspapers on the floor and dump the trash onto the newspaper.

Have your classmates calculate how much trash is generated by their families each day. Log the answers in your results.

Research and find out how many people live in your state. Calculate an average of how much trash is produced each day by people in your state. Log the answer in your results.

Discuss where all this trash goes. Explain that landfills take up space that could be used for agriculture or wildlife. Ask what other negative effects landfills might have on the environment (air pollution from toxic gases, poisonous drainage into groundwater).

Calculate the amount and type of trash your class throws out each day in the classroom. As each of your classmates throws something in the trash container, log it on a sheet of paper. At the end of the day, weigh the trash your class threw away. (The trash might be too heavy for you to lift. If so, ask your teacher or a school maintenance person to weigh it for you.) Record the weight in your results. Find out how many classrooms there are in your school, school district, and city, and calculate how much trash is thrown out daily. Calculate an average of how much trash your class throws away each school year. Log it. Where does all this trash go? Research sanitary landfills. Are they better than other kinds of landfills?

Obtain permission from your teacher, school principal, or lunchroom supervisor to calculate and analyze the trash thrown away each lunch hour. Position yourself in the lunchroom by a large trash can. As the students throw out what is left after they have eaten lunch, log the items. After the lunch hour, weigh the trash container. If there is more than one container, weigh each one. Record your calculations in the results.

Go over the analysis of the trash with your classmates. Ask: "Is there anything else we can do with these items we throw away?" On paper, organize the trash into the following categories: items for reuse, items for recycling, and items for a landfill.

Discuss where your class can take items to be recycled. Perhaps you could start a class recycling project.

List on paper all the reasons for reduction and recycling of waste. Below this, list ways in which each individual can cut down on the amount of solid waste he or she produces. Record on your paper any suggestions made by classmates. Make a copy of this paper for each member of your class.

Finally, ask how many of your classmates now believe that recycling is beneficial to the environment. Record your results.

***Results.*** 1. Log of family trash weight. 2. Log of trash generated by state. 3. Log of weight of trash thrown out in classroom in one day. 4. Log of weight of one day's trash from school, school district, city schools. 5. Log of weight of trash thrown out yearly by classroom. 6. Log of type of trash thrown out in school lunchroom. 7. Log of weight of lunchroom trash for one day. 8. Reasons for reduction and recycling. 9. Ideas for reduction of waste. 10. Vote on reduction and recycling awareness.

## OUR PLANET'S RESOURCES

Many of the products we use in our everyday life come from the planet's natural resources. A natural resource is a substance that nature has produced, such as air, water, minerals, trees, coal, and oil.

Natural resources provide the raw materials for many of our products. A *renewable* natural resource is one that is taken from an incredibly vast or self-renewing source. However, few things in nature are truly infinite. A natural resource is only renewable if it is managed properly and not abused. When they are managed wisely, these resources are replaced naturally or through human-assisted systems. Some good examples of renewable resources are the world's forests. If the trees are cut down according to a well-thought-out plan that allows much of the forest to still stand, and if people plant seedlings to replace the trees taken out, a forest could last indefinitely.

A *nonrenewable* resource is a naturally occurring substance that is available in only limited amounts. It may take many years for nature to produce these resources, and therefore they are essentially nonreplaceable. *Fossil fuels*—coal, oil, and natural gas—are considered to be a nonrenewable natural resource because they take millions of years to form. They cannot be replaced within any realistic period of time. Thus, once they are used up, they are gone for good.

In a way, trash is both renewable and nonrenewable. Trash contains many materials that can be reused or recycled. It also contains materials that cannot be put to any useful purpose again.

The proper management of both trash and our natural resources will ensure a clean and healthy environment. Conservation of natural resources and reduction of trash are key to our continued success on this planet.

## THE MAKEUP OF SOIL

One of our greatest natural resources is the soil. If you were to examine closely a handful of soil, you would see a world teeming with tiny forms of life. This world sustains the rest of the living environment. The activities of the *microorganisms, microbes* for short *(bacteria, protozoans, fungi),* mites, springtails, earthworms, and other creatures that inhabit the soil aid in the decomposition of living and nonliving things. Through the activities of these organisms, nutrients are made available to plants. This is a vital part of the food web. Without it, life on Earth could not exist.

Examining the soil, you can see that it is anything but solid. Soil contains many spaces between its particles, spaces that are filled with air, water, and living organisms. Life in the soil exists between the particles of soil, not in it.

If you were to drill into the ground, you would see that soil is made up of six different layers, each with its own distinct properties. The very top layer is referred to as the organic layer. Under this layer lies a layer of accumulated organic matter called topsoil, or humus. Beneath this is the eluvial layer, which consists of material deposited from other areas by erosion. Next is the subsoil, a layer of mineral-laden clay. Then comes the rocky debris layer, which consists of bedrock that is somewhat broken down by physical and chemical processes. Finally is the solid bedrock, the base layer. This is the source of the soil itself.

It takes many years to produce the rich, life-giving soil described above. However, with poor management and erosion, that soil can disappear quickly. Without soil, the Earth would be a barren planet, incapable of sustaining life.

## Science Project #3— Life Within the Soil

### *Materials Needed*

- shovel or hand spade
- newspaper
- magnifying glass or stereo microscope
- small piece of window screening or cotton gauze
- berlese funnel (funnel with filter)
- ring stand with ring
- jar
- alcohol
- 100-watt light bulb and socket
- medicine dropper
- blank slides and cover slips
- compound microscope

***Inference.*** A cubic foot of soil will contain a variety of life.

***Hypothesis.*** If a cubic foot of soil is studied, it will reveal a variety of living organisms.

***Procedure.*** Collect a few cubic feet of soil samples from several different sources—for example, your backyard, near a stream or river, and from a woodlot. Place each sample on a sheet of newspaper, and examine it with the naked eye and then with a magnifying glass. What do you see? Write down your observations in your results.

Set up a drying and collection station. Place a small piece of window screen or cotton gauze inside the funnel, with a sample of soil on top of the screen. Set up your ring stand and place the funnel in the ring. The funnel's tip should extend down into the mouth of a small jar filled with

¼ inch of alcohol. Set up a 100-watt light bulb over the opening of the funnel to dry out the soil. This will cause the organisms to move out of the soil and into the collection jar.

Observe some of the organisms you have collected. You can see many of them with a magnifying glass. Use a medicine dropper and place a drop of alcohol from the jar beneath the funnel on the slide. Cover it with a cover slip. Put the slide on the stage of a compound microscope, and examine it for organisms. (You may want to ask your science teacher to help you with the compound microscope.)

How many kinds of organisms did you collect? List them in the results and diagram some of them.

***Results.*** 1. Organisms seen with magnifying glass. 2. Organisms seen under the microscope. 3. Number of species collected. 4. Diagrams.

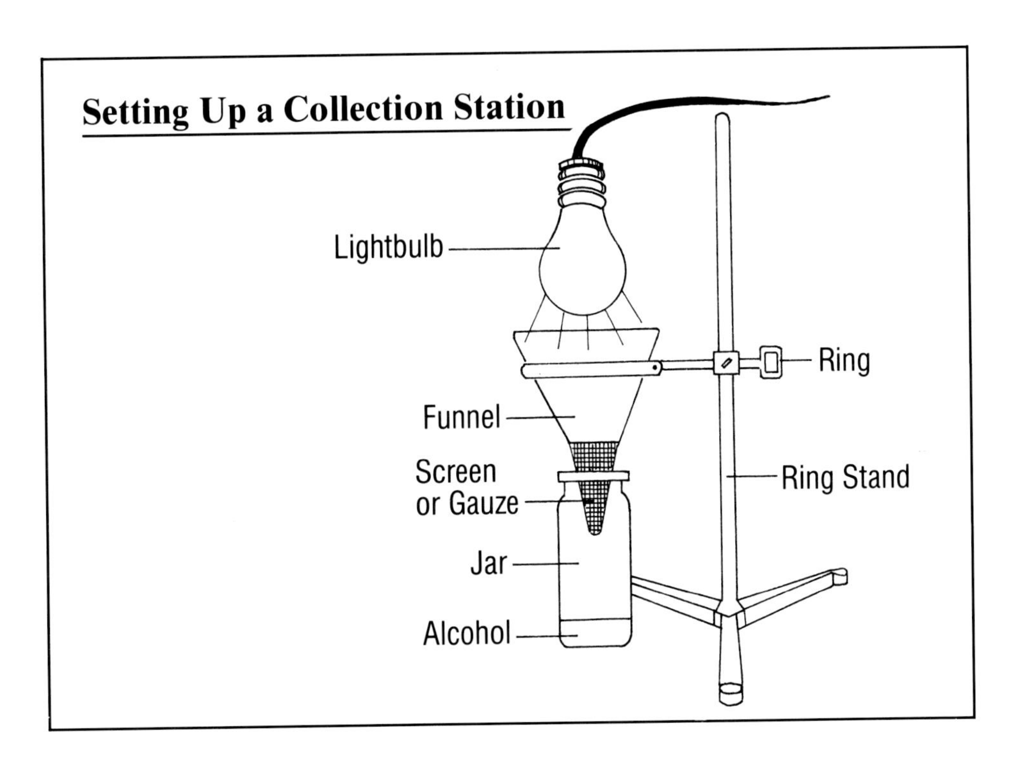

## Organisms Seen Under a Microscope

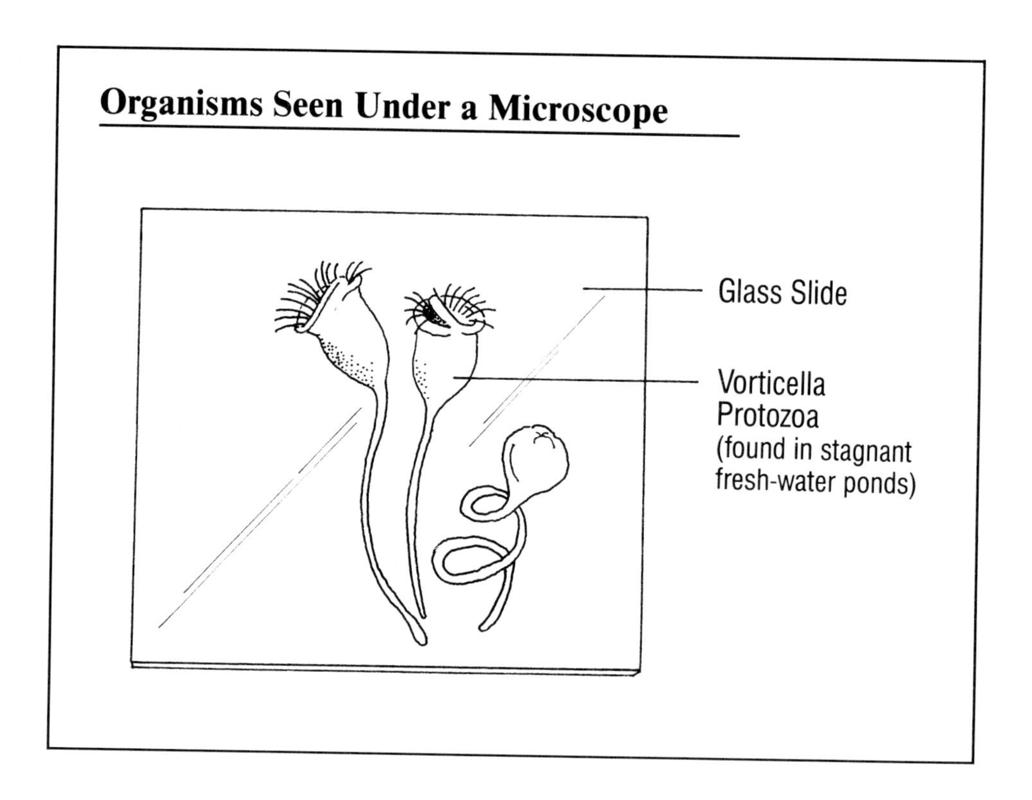


## NATURE'S CONSERVATIONISTS

Among the lower animals, the earthworm is considered the agriculturalist. It is always busy plowing and fertilizing the soil. Burrowing deep into the ground, sometimes as deep as 8 feet but most often from 12 to 18 inches, earthworms bring the subsoil to the surface, thus renewing the soil itself.

To break up the soil, the earthworm has a gizzard filled with grains of sand or fine gravel. As the worm takes in soil, it grinds it in its gizzard, making it much finer. It then deposits this ground soil in the top layers of the earth, leaving a wonderfully nutritious medium in which plants can grow.

As the worm burrows, it drags into its tunnels leaves, flowers, grasses, other yard debris, as well as shells and the bones of dead animals. These organic materials decay and thus furnish food for plants. Worm burrows also provide excellent drainage for the soil, allowing water to work its way down into the lower levels.

*The earthworm is one of nature's most talented tillers of the soil.*

The earthworm's habitat is damp to moderately wet, fine soils. Earthworms need moisture, since they get their oxygen for breathing from the water. They breathe through their skin. The worm is nocturnal (active at night). During the day, it lies in its burrow near the surface, head toward the outside world, its body extended to its full length. This makes it easy prey for birds and other creatures who consider the earthworm a delectable meal. The earthworm is omnivorous, eating a variety of things, such as earth, leaves, flowers, raw meat, fat, and sometimes even other earthworms.

## Science Project #4—
## An Earthworm Farm

### *Materials Needed*

- 5- or 10-gallon aquarium
- soil from your yard or a woodlot (fill the tank half full)
- 24 night crawlers (can be obtained from most fish bait stores)
- grass clippings and dead leaves
- dark construction paper or oilcloth
- duct tape
- magnifying glass
- encyclopedia or other reference book
- small flashlight

***Prediction.*** In an earthworm farm, earthworms will burrow. In the process, they will refine the soil and establish drainage.

***Hypothesis.*** If an earthworm lives in the soil, then it will utilize and improve that soil.

***Procedure.*** Set up an earthworm farm by filling a glass aquarium half full of soil from a yard or a woodlot. It is important to have this natural soil, because it contains debris and organisms that cannot be found in packaged soil. Add the night crawlers to the soil, and cover the soil with a thin layer of grass clippings and dead leaves. Cover the outside of the aquarium with dark construction paper or oilcloth. Attach the paper or

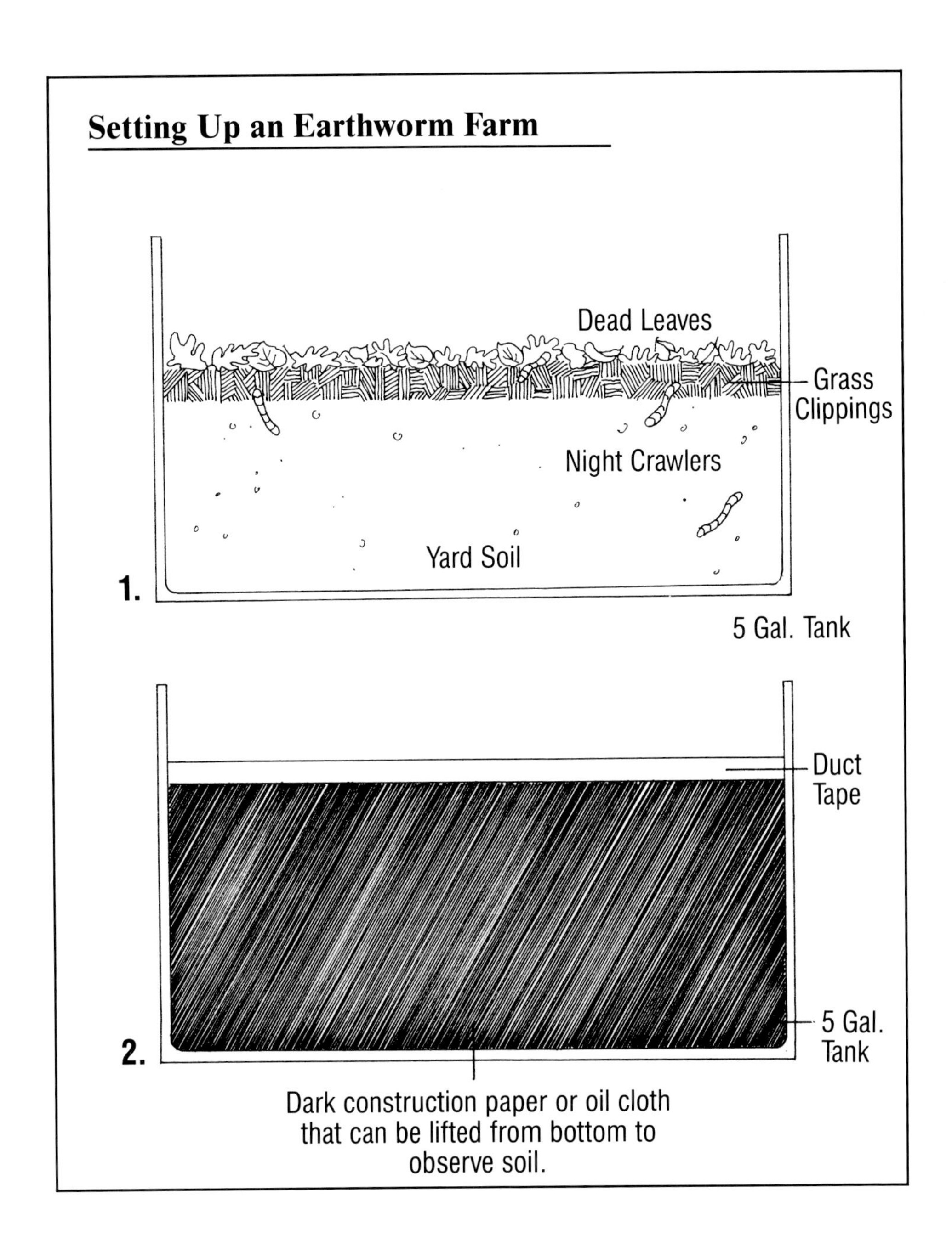
Setting Up an Earthworm Farm
Dead Leaves
Grass Clippings
Night Crawlers
Yard Soil
1.
5 Gal. Tank
Duct Tape
2.
5 Gal. Tank
Dark construction paper or oil cloth that can be lifted from bottom to observe soil.

cloth to the sides of the tank with duct tape. Attach it only at the top, so that it can be lifted from the bottom to observe the soil.

Allow the aquarium to stand for two weeks, and then begin your observations.

Remove an earthworm from the tank, place it on some soil on a table, and observe it. How does the worm crawl? Has it any legs? Look at the worm with a magnifying glass. Can you see now how it moves? List the answers in your results.

Diagram the earthworm's body. Draw the proper shape and color it correctly. Label its parts. (Use an encyclopedia or other reference book for help.)

Lift the flaps on the aquarium and observe the soil for burrows. Describe them. Is any burrow occupied by more than one worm? Watch the worm move and eat in its burrow. Describe this.

Observe the aquarium at night, in the dark. Use a small flashlight. Where are the worms at night?

Is the earthworm a good tiller of the soil? How can you tell?

***Results.*** Answers to questions and diagram.

## THE SANITARY LANDFILL

A sanitary landfill is a dump where the waste is crushed into layers and each layer is covered with soil. Bacteria, mold, yeast, protozoans, and moisture will then supposedly decompose the trash. This is not necessarily a true reflection of what happens, however, since not everything decomposes. For instance, glass does not decompose. Also, items that do decompose may do so at different rates. Because of the compacting effect of the layers of soil, not enough air gets to the trash to allow the bacteria to do their work. Items that (at least in theory) will decompose in landfills are said to be biodegradable.

## Science Project #5—
## Which Items Will Decompose in a Landfill?

***Materials Needed***

- large wooden or cardboard box, approximately 24″ by 36″ by 18″
- heavy plastic to line box (can be large garbage or lawn bags cut open)
- tape (duct tape or masking tape)
- soil (most preferably soil dug from a backyard or some other outside source. This soil is preferred because it will contain the microbes necessary for decomposition. Soil that is bought in bags is often sterilized and free of most microbes.)
- hand spade
- materials to be landfilled: scraps of wood, strips of rubber, newspapers, toilet paper, plastic wrap, aluminum can, tin can, brown paper, a penny, an iron nail, pieces of hard plastic (cut up a 2-liter soda bottle), bits of cloth (cotton, wool, polyester) from fabric remnants, Styrofoam™ nuggets (used in packaging), food scraps (bread, meat, vegetables), aluminum foil, and glass items (DO NOT use broken glass)
- paper and pencil
- water in a sprinkling can

***Inference.*** Not all items in a landfill decompose.

***Hypothesis.*** If unsorted trash is put into a landfill, then much of it will decompose, but some items will not decompose.

***Procedure.*** Line your wooden or cardboard box with your plastic liner. Pull the plastic over the rim of the box and tape it on the outside to ensure that it remains in place. Now spread about 6 inches of soil over the bottom of the box. This done, place your trash samples in the soil. Make a diagram of the box and label where you put each trash sample.

## Testing Items in a Landfill

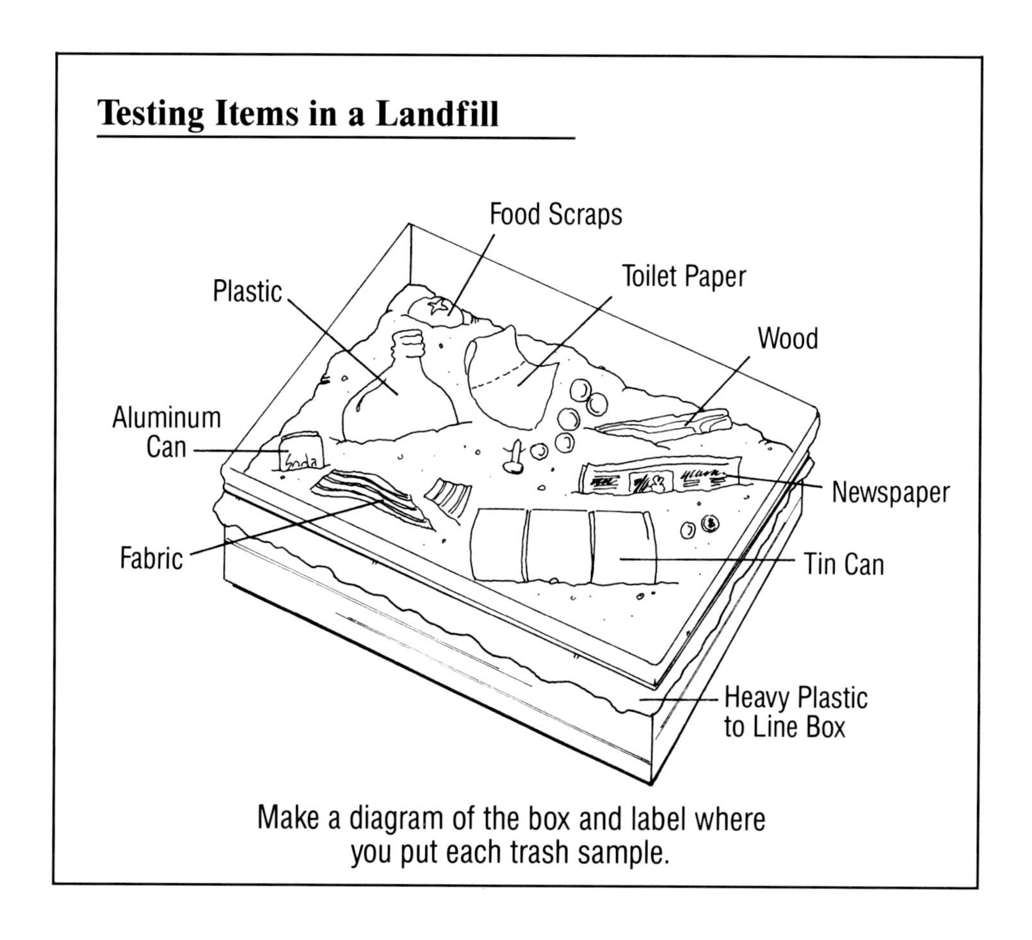


Make a diagram of the box and label where you put each trash sample.

Cover the trash samples with another 6 inches of soil. Sprinkle the soil with enough water to make it moist. DO NOT SOAK THE SOIL. It should be barely moist.

Set the box aside for one month and let the microbes do their work. After a month has passed, dig up the wood samples and see if enough decomposition has taken place for you to draw some conclusions. If not, bury them again and let the box stand for another month.

After the items have started decomposing, examine each item and compare the rate of breakdown. List your comparisons on a chart.

## Chart of Results

| | 1 Month | 2 Months |
|---|---|---|
| | | |
| wood | | |
| rubber | | |
| newspaper | | |
| toilet paper | | |
| plastic wrap | | |
| aluminum can | | |
| tin can | | |
| brown paper | | |
| penny | | |
| iron nail | | |
| hard plastic | | |
| fabric | | |
| styrofoam | | |
| food scraps | | |
| aluminum foil | | |
| glass | | |
| | | |
| | | |

(MAKE YOUR OWN CHART OR PHOTOCOPY THIS ONE.)

## THE GROUNDWATER PROBLEM

Water is found on Earth in many forms. *Groundwater* is water beneath the surface of the Earth. There are underground lakes, rivers, and streams. Usually, when water percolates through a landfill, the toxic seepage, or *leachate,* goes into the groundwater and spreads from there to other sources of water. Water that is contaminated by the seepage from a landfill is usually unfit for humans to drink and is also harmful to the aquatic organisms that live in this environment.

### Science Project #6—
### Water Pollution from Landfills

***Materials Needed***

- 10- or 20-gallon fish tank (preferably the larger one)
- sponge
- abrasive cloth
- vacuum cleaner
- piece of glass cut to fit the width of the tank
- silicone sealant
- caulking gun
- framed screen cut to fit ⅔ of the tank
- soil from an outdoor source, such as a garden or backyard
- hand spade
- scraps of garbage (wood, paper, foodstuffs, etc.)
- gravel
- water pitcher
- pH test paper
- LaMont Chemical Kit (can be obtained by writing Spectrum Educational Supplies Ltd., 9 Dohme Avenue, Toronto, Ontario, Canada)
- thermometer
- 2 saucers
- medicine dropper
- plastic wrap
- unflavored gelatin
- compound microscope
- blank slides and cover slips
- biology reference book

***Prediction.*** A leaky landfill will most likely contaminate groundwater.

***Hypothesis.*** If a landfill is located directly over a groundwater source, then its seepage will contaminate the groundwater.

***Procedure.*** Clean the fish tank thoroughly, using a sponge and an abrasive cloth. DO NOT use soap. Soap will remain in the tank and alter the results of your experiments. Remove all the dust and particles from the tank by vacuuming the bottom and corners of the tank.

With the tank clean, you are ready to construct the pond, underground water system, and landfill area. In order to separate the pond from the land area, a piece of glass should be inserted into the tank. The glass should be positioned so that one third of the tank is the pond. Secure the glass in place by running a bead of silicone sealant from the caulking gun along the bottom and sides of the tank. Carefully place the glass into

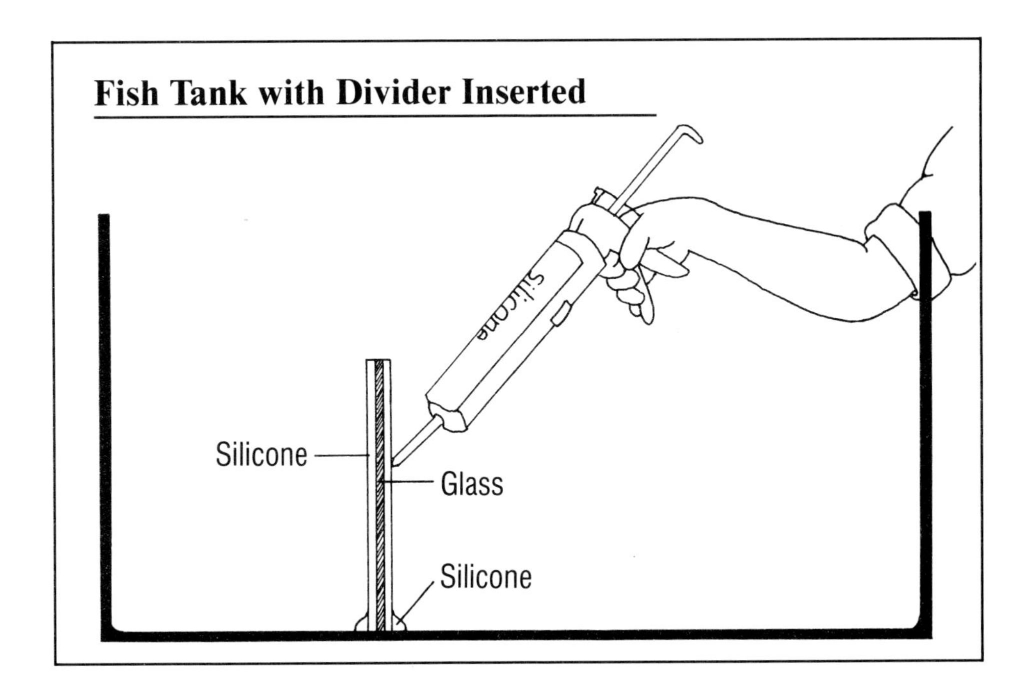

the silicone bead. With your finger, carefully spread the silicone along the seam where the glass meets the sides and bottom of the tank. The silicone, which takes twenty-four hours to cure, will form a watertight seal. It will prevent seepage from one section of the tank to another.

After the silicone has cured, position your framed screen in the bottom of the tank. This screen may be bought at a hardware store or made yourself with screening stapled or tacked to pieces of wood. The screen will take up two thirds of the tank bottom not used for the pond. Make sure the screen fits the bottom securely. In order to make sure the screen is not touching the bottom, tack 3-inch-high pieces of wood to the underside of the screen on both ends. Silicone the screen in place to prevent movement and the accidental contamination of the groundwater area.

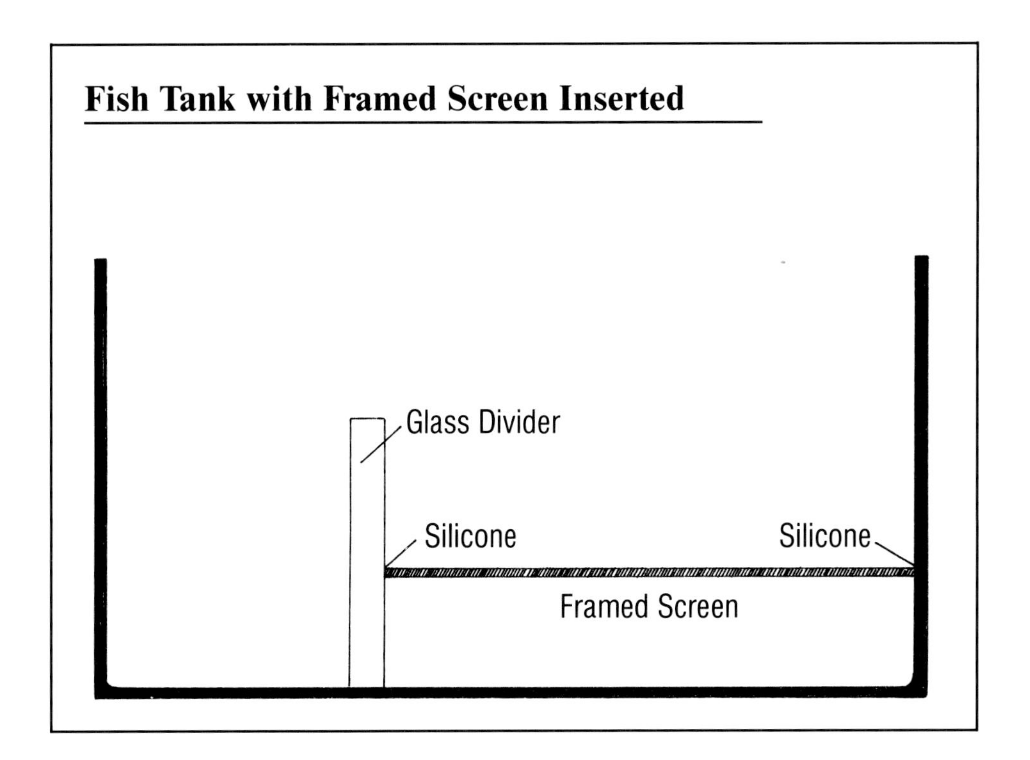

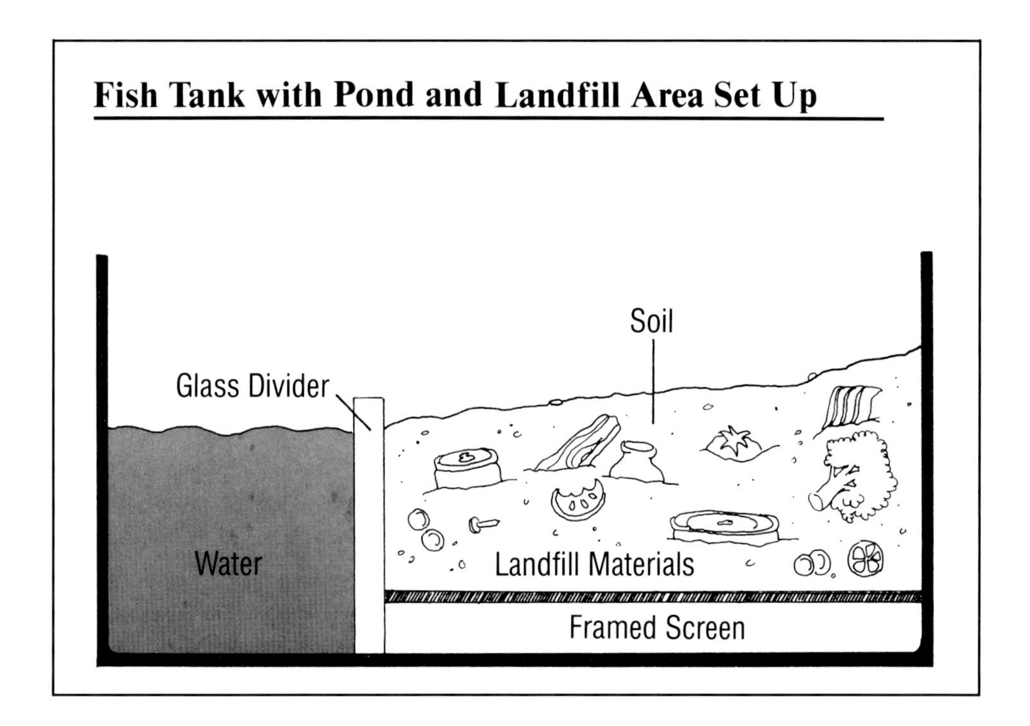


Cover the bottom of the pond area with an inch or two of gravel. Cover the screened-bottom area with soil. Put in approximately 6 inches of soil. Add your scraps of trash and garbage to the soil. Use the same trash materials as in Science Project #5, but add a lot more food scraps. Cover the trash samples with enough soil to reach just below the top of the glass divider (approximately 6 inches).

Fill the pond with water to just below the top of the glass, and sprinkle water on the landfill area of the tank.

Allow the landfill and pond to stand for at least two months, or until decomposition is well under way. (You will estimate the time of decomposition better if you completed Science Project #5 first.) Sprinkle the landfill area with water each day. If the groundwater area, under the screen, fills with water, stop sprinkling for a day or so.

Once your materials have started to decompose, you are ready to start testing the groundwater and the pond water. Test the pond water first.

The pH scale measures the level of acidity or alkalinity of water. An *acid* is any substance that has a sour taste, turns blue litmus paper red, measures 1 to 6 on the pH scale, usually contains hydrogen (H), and is corrosive (wears away other substances). A *base* is any alkaline substance that has a bitter taste, turns red litmus paper blue, measures 8 to 14 on the pH scale, usually contains a hydroxyl molecule (OH), and is slippery to the touch.

Use the pH paper to test first the pond water then the groundwater for acidity or alkalinity. An acid will register 1 to 6 on the pH scale; a strong acid will register 1 to 3, a weak acid 4 to 6. A measurement of 7 is neutral. Numbers 8 to 14 indicate a base (alkaline substance); 8 to 10 indicate a weak base, while 11 to 14 indicate a strong base. Record your results.

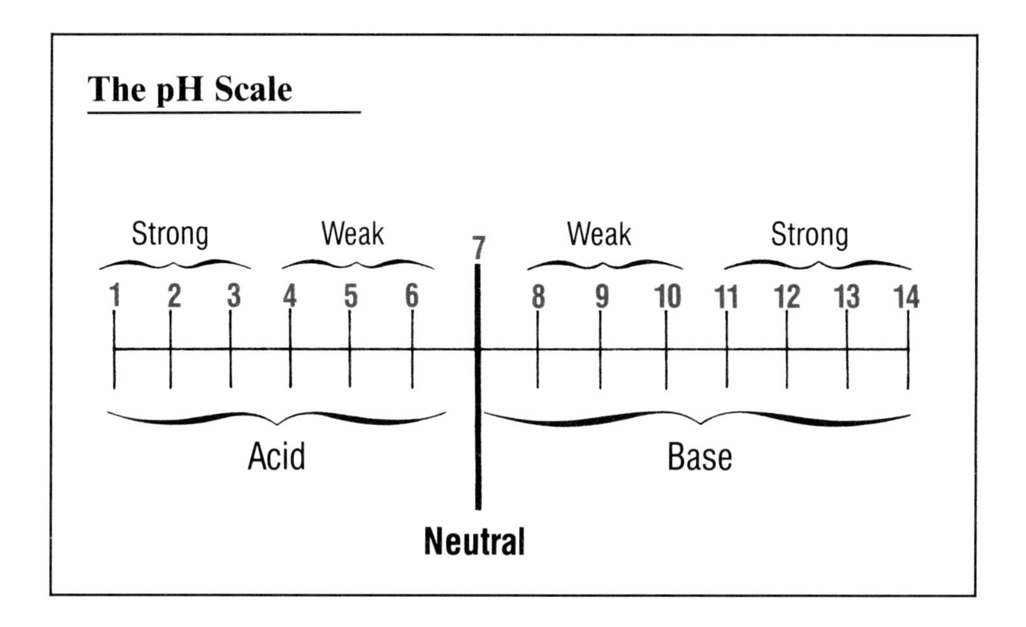

Use the LaMont Chemical Kit to test the pond water and groundwater for dissolved oxygen. A lack of dissolved oxygen in water is a valuable pollution indicator. The respiration of aquatic plants at night (plants release carbon dioxide at night), the aluminum and iron from decomposing metals, the carbon from decomposing organic matter—all of these use up the oxygen in water. The less oxygen water contains, the more likely it is polluted.

Test your pond water and groundwater for iron, aluminum, carbon, sulfur, and lead. (If you write the chemical company that these are the chemicals and gases that you want to test for, they will send you the proper materials to conduct the tests.) Record the results.

Also take the temperature of your water. Changes in temperature can indicate an increase of decomposing matter. Water with a higher temperature usually holds less dissolved oxygen than cooler water. Record your results.

Finally, check your water for bacteria. Bacteria are microbes that can be seen only under a microscope. There are many helpful varieties of bacteria. However, other bacteria cause disease and can be extremely dangerous if they are in a community's water supply.

Test for bacteria by half filling two clean saucers with gelatin prepared according to package directions. Allow the gelatin to set. Using the medicine dropper, remove some groundwater from the tank and place it in one of the saucers of gelatin. Label this saucer "*A*." Thoroughly rinse the medicine dropper, then remove water from the pond area of your tank and put it in the other saucer of gelatin. Label this saucer "*B*." Cover each saucer with clear plastic wrap to keep out dust and mold from the air. Observe the saucers daily and note the changes that take place. Gray, green, black, or cloudy areas that appear in the gelatin are colonies of bacteria. (Be careful not to touch or ingest any of the gelatin at this point.) Did bacteria appear in both saucers? Record your results.

Once the bacteria have developed, have your science teacher or some other adult who knows how to use a microscope help you to place some of the bacteria on a blank slide. Put the slide on the stage of a compound

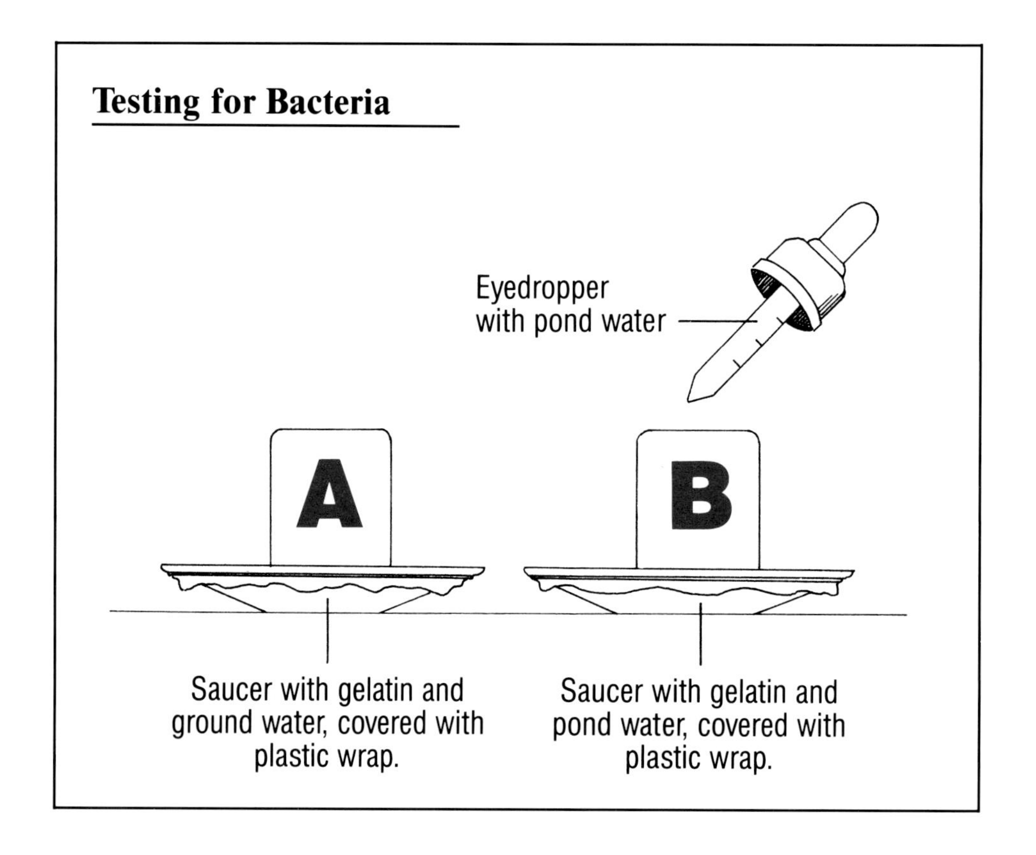


microscope and focus. Diagram what you see. Then, using a biology textbook or reference book, look up the kinds of bacteria you are observing. What diseases are these bacteria associated with? Record results.

***Results.*** 1. Record of the pH of the pond water and the groundwater. 2. Log of dissolved oxygen in pond water and groundwater. 3. Log of chemicals in pond water and groundwater. 4. Log of temperature of pond water and groundwater. 5. Results of bacteria test in saucers *A* and *B*. 6. Diagram of bacteria seen under microscope and list of the diseases they cause.

# 4

## *Nature's—and People's— Recycling Efforts*

### THE NITROGEN CYCLE

Nature, in order to exchange materials between its living and nonliving aspects, has had to develop its own recycling program. Because nature's cycles involve living things and the exchange of chemicals, these cycles are referred to as *biochemical cycles*. All of nature's cycles involve water and gases or soils.

An important element that cycles in this way is nitrogen. The *nitrogen cycle* is essential to all organisms. Nitrogen helps organisms to manufacture proteins, which are made of amino acids and which help build tissue in plants and animals. Nitrogen is the most plentiful gas in the air. But, as a gas, it is unusable by most plants and therefore cannot be cycled through the food web. In order to be used by plants, nitrogen must be in the form of nitrates, compounds made mostly of nitrogen and oxygen. Plants absorb these nitrates through their roots and use them to aid in their growth and tissue development.

Once the plants have absorbed the nitrates, the nitrogen is transferred to other organisms through the food web. Herbivores eat the plants; carnivores eat the herbivores; and omnivores eat the plants, herbivores, and carnivores. At each level, nitrogen is absorbed and used to build tissue.

*As part of the nitrogen cycle, plants absorb nutrients through their roots and use these nutrients to live and grow.*

All plants and animals create waste and eventually die. Waste and decaying matter are broken down by decomposers. For example, some bacteria change waste and decaying matter into amino acids. Different bacteria change these amino acids into ammonia, a nitrogen compound. Still different bacteria change the ammonia back to nitrites. The nitrites are converted by other bacteria into nitrates. These nitrates are reabsorbed by the plants and used to make more living tissue.

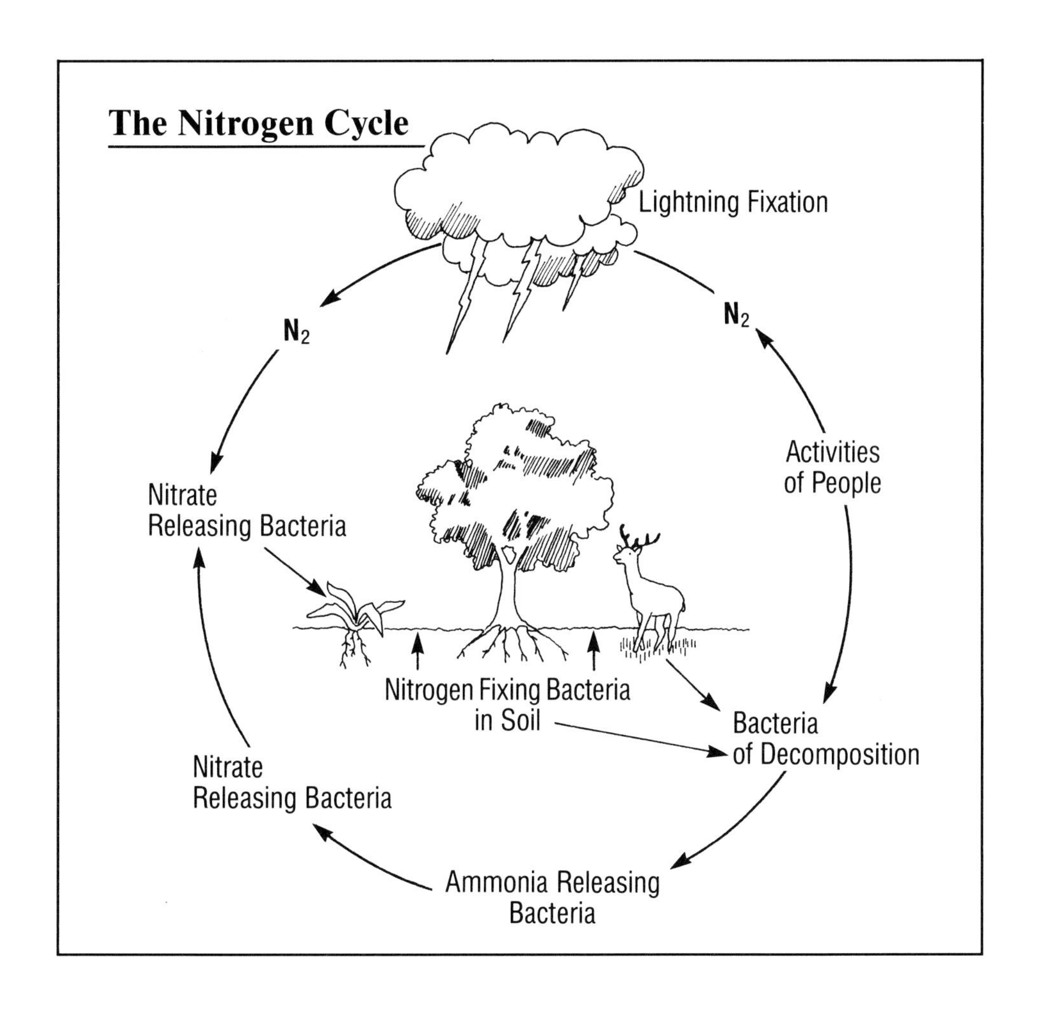

## THE OXYGEN CYCLE

Another important natural cycle is the *oxygen cycle*. Oxygen is found dissolved in water. It also makes up 21 percent of the atmosphere. Oxygen is produced by green plants as a by-product of photosynthesis, the process green plants use to make food. Most living things use this oxygen, returning it to the air and water in the form of carbon dioxide. The carbon dioxide is then used by green plants in photosynthesis. This process once again replenishes the oxygen in the air by releasing it as a by-product.

### Science Project #7— The Oxygen Cycle: The Star of Nature's Recycling Program

***Materials Needed***

- 2 test tubes
- 2 elodea plants (may be obtained from most pet stores that carry fish)
- 2 corks to fit the test tubes
- soda lime crystals (to absorb carbon dioxide)
- test tube rack

***Inference.*** Plants produce oxygen. This is used by animals, which produce carbon dioxide, which the plants then use to produce more oxygen.

***Hypothesis.*** If green plants are to produce oxygen through the process of photosynthesis, then they must have carbon dioxide to complete the cycle.

***Procedure.*** Fill a test tube with water, and put an elodea plant in it. Seal the test tube with a cork. Label this test tube "*A*" and place it in a test tube rack. Fill a second test tube with water. Add about ¼ teaspoon of soda lime crystals to the water. Add an elodea plant to the water. Seal the test tube with a cork. Label this test tube "*B*," and place it in the same test tube rack. Put the test tube rack in an area that receives a great deal of sunlight, preferably a window with a southern exposure.

Observe the plants carefully for the next two weeks. Note the elodea plant producing oxygen. (You will be able to see the oxygen bubbles forming on the plant's leaves and then rising in the water.) Keep a chart of what happens in test tubes *A* and *B*.

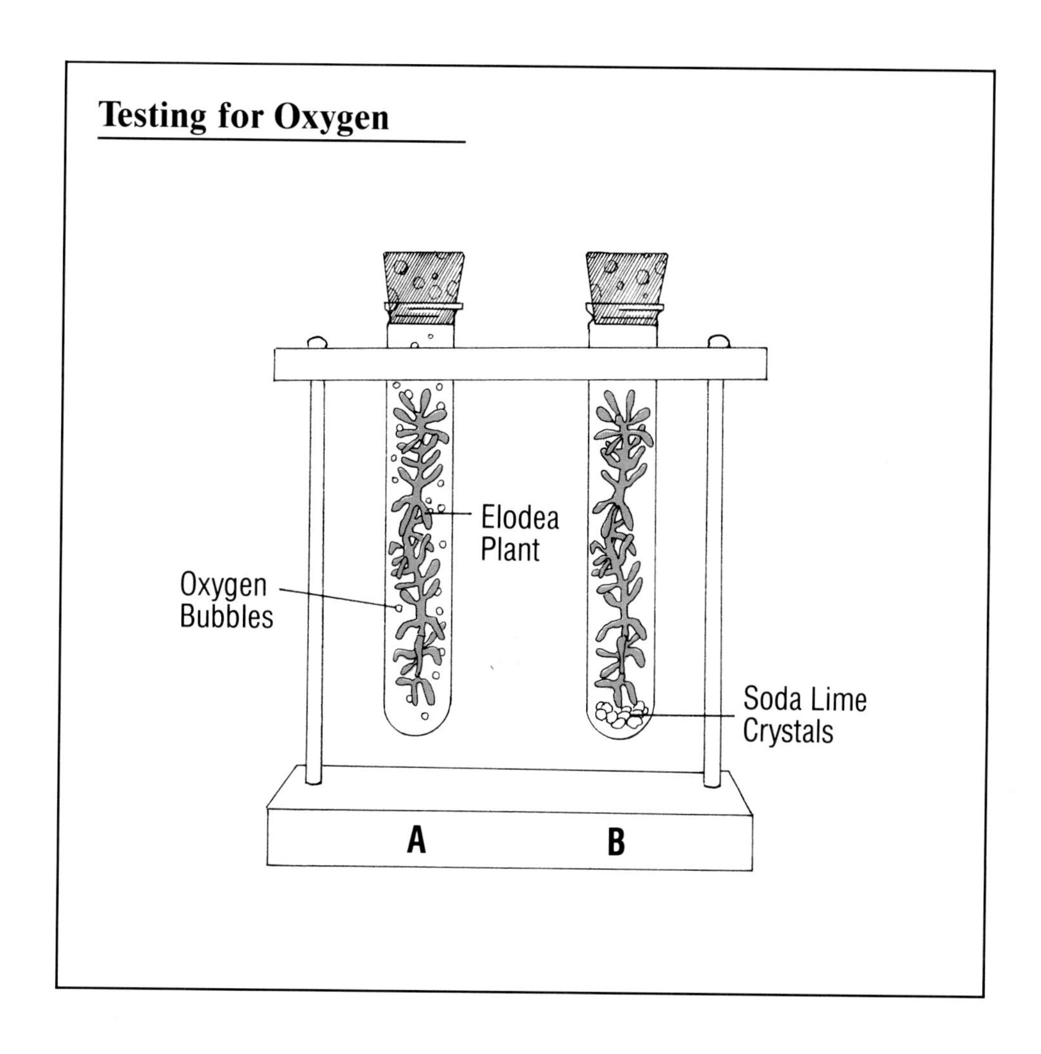

## THE RECYCLING OF HUMAN-MADE PRODUCTS

Like nature, people can recycle materials. Recycling reduces waste and conserves our natural resources. By reducing and reusing waste, we also help to avoid a worldwide garbage crisis.

Microbes are tiny plants and animals that aid in the process of decomposition. Microbes such as molds, bacteria, yeasts, and protozoans turn decomposing matter into nutrients, which are then absorbed by plants. As we saw earlier, earthworms also take part in the process of decomposition, and so do some insects.

However, microbes are the most efficient decomposers. An earthworm digests its own weight in food daily. A microbe digests its own weight in food in seconds!

Microbes such as molds and bacteria exist in colonies. These colonies consist of millions of individual cells or organisms. The microbes digest and oxidize (burn) garbage. This results in a good, rich humus that enhances the fertility of soils.

A *compost heap* is really a busy complex of microbes and other decomposers. Millions of bacteria are the first organisms to go to work on organic (living or once living) tissue. Later, they are joined in the composting process by molds and protozoans, which continue the job. Eventually, bacteria and molds are joined by beetles, centipedes, millipedes, and earthworms. All work together to break down organic matter.

Up to twenty percent of the household garbage we throw out can be recycled through composting. Throwaways such as fruit pits, vegetable peelings, eggshells, coffee grounds and filters, leaves, grass clippings, and other biodegradable, organic wastes can be composted. However, DO NOT COMPOST MEAT, BONES, SALAD OILS, OR OTHER FATTY SUBSTANCES. These items attract pests and slow down the process of decomposition.

By mixing compost with soil in your yard or garden, you return organic matter to the soil in a usable form. The decomposition of organic matter is essential for the continuation of life on Earth. Decaying organic matter improves plant growth by loosening the soil, improving the capacity of the soil to hold water, and adding essential nutrients to the soil.

By composting, we can reduce the amount of garbage sent to landfills. We can also reduce problems that organic matter can cause in landfills. Yard wastes and other organic matter produce methane gas in landfills. They also create a strong leachate, which can then flow into groundwater and contaminate it.

## Science Project #8— Making Compost at Home

### *Materials Needed*

- tools to build bin
- ready-mix concrete
- 35 feet of snow fencing
- 6 pieces of wood for fence posts, 2″ by 4″ by 4′
- wire to fasten fencing to wood
- small branches or other coarse material
- kitchen scraps (no meat or fatty substances)
- leaves
- grass clippings
- sawdust
- plant clippings
- water
- soil
- manure
- lime or wood ash
- straw
- large thermometer
- 3′ wooden dowel
- pitchfork

***Inference.*** Composting will reduce the amount of garbage generated by a household.

***Hypothesis.*** If organic household wastes are composted, then the amount of garbage sent to be landfilled will be reduced.

***Procedure.*** Monitor your household's garbage daily for one month. List the materials you throw out, including grass clippings, leaves, and other yard wastes. Weigh daily as much of the garbage as you can. Record this weight in your results.

Decide on an area in your backyard where you would like to put your compost heap. The heap should be in a sunny spot, but in an area that is not used for recreation and where it will not be an eyesore.

The heap described in this book will be 5 feet square, with a turning bin 5 feet square right next to it. (What you have is actually one large bin divided in the middle.)

First, remove any grass or sod covering the area where you have chosen to build your compost pile. This is important. In order for the compost pile to function properly, it must have direct contact with the soil. This will allow the microbes in the soil to enter the compost heap.

With the sod removed, dig a hole 1 foot deep in each corner of your 5-foot square, and fill it with ready-mix cement. (If you do not want the fence to be permanent, place a post in each hole and pack the hole with soil. The tightly packed soil should hold the post in place.)

Once your fence posts are secure, wire the snow fence to the posts. The section where the last of the fence attaches to the fourth post can be used as a gate.

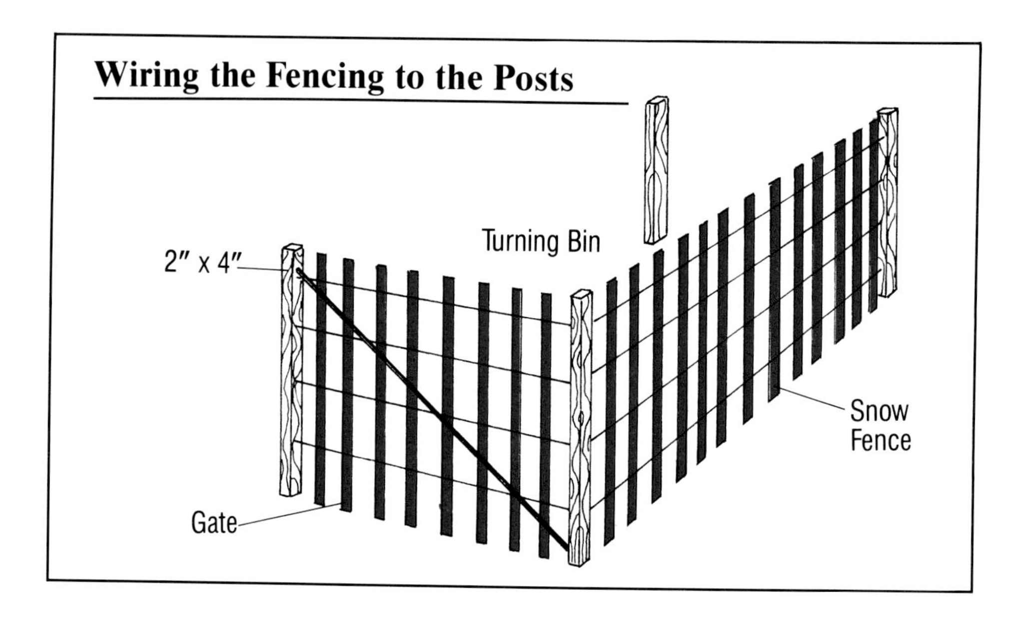

The bottom layer of the compost pile should consist of 3 to 4 inches of brush or wood chips, which should be chopped or broken up into small bits. This coarse material will provide excellent drainage for the heap. It will also allow for good air circulation around the base of the pile.

Lay 6 to 8 inches of mixed kitchen scraps (see materials list), leaves, grass clippings, yard wastes, and sawdust over the base of the compost heap. Moisten this layer. It should feel like a dampened sponge.

The third layer of the compost heap should consist of at least 1 inch of yard soil. This layer will add necessary microorganisms to the pile. Remember, the microbes are the most important ingredient of the heap. Without them, the heap will not function properly.

Manure will provide the nitrogen needed by the microbes. The fourth layer should consist of manure with lime or wood ash sprinkled over the top of it. The lime or wood ash will reduce the acidity of the heap. This layer should also be damp.

Keep making layers in this order until the bin itself is full. Spread straw over the top of the completed heap, to prevent rain from washing away the compost's nutrients.

Allow the heap to stand for a week and then check its internal temperature. Do this by attaching a thermometer to the 3-foot dowel. Stick the thermometer into the heap's center, and let it stay there for a minimum of 15 minutes. A well-functioning heap will have an internal temperature of 140 to 160 degrees Fahrenheit.

Allow the heap to stand for six weeks and then turn it. Do this by forking the heap into a new pile in your turning bin. Fork the outside of the old pile into the center of the new pile. Moisten the heap if it needs it. It should be damp, like a squeezed-out sponge. Cover the turned heap with straw and wait approximately four months. (If you turn the pile every few weeks, the compost may be ready sooner.) At this time, your heap should be ready to use. If you start your heap in autumn, it will be ready in spring.

The dark brown compost should be mixed into the top 3 inches of soil in your yard or garden. Mix about 3 cups compost to every square foot of soil.

## Layering of the Compost Pile

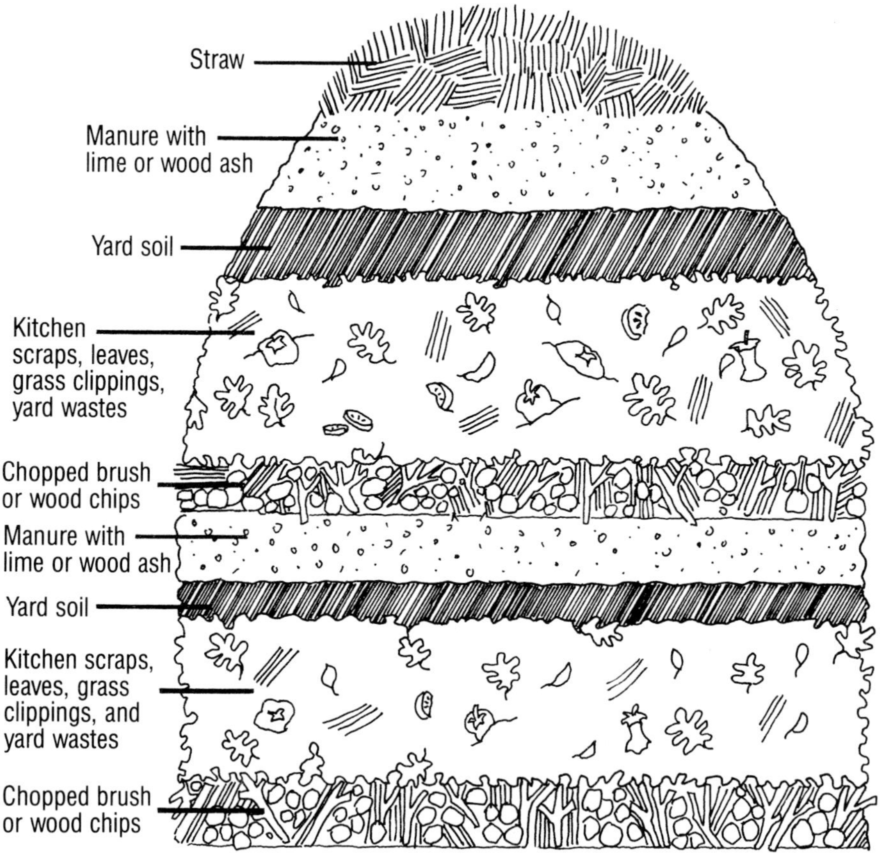

Straw

Manure with lime or wood ash

Yard soil

Kitchen scraps, leaves, grass clippings, yard wastes

Chopped brush or wood chips

Manure with lime or wood ash

Yard soil

Kitchen scraps, leaves, grass clippings, and yard wastes

Chopped brush or wood chips

Once you have started composting your kitchen and yard wastes, continue to monitor your household garbage as described at the beginning of the procedure for this project. Record the results.

## Chart of Results

| | Before Composting | | After Composting | |
|---|---|---|---|---|
| average daily weight | | | | |
| average weekly weight | | | | |
| Materials and percent of Garbage | Food scraps | % | | % |
| | Grass clippings | % | | % |
| | Leaves | % | | % |
| | Yard wastes | % | | |
| | Paper | % | | |
| | | | | |
| | | | | |
| | | | | |
| | | | | |
| | | | | |

(MAKE YOUR OWN CHART OR PHOTOCOPY THIS ONE.)

# 5

# *Energy and Garbage*

It takes enormous amounts of energy to run modern societies. Most of our energy needs are met by fossil fuels. Aside from the fuels used for transportation, electricity, and home heating, our energy supplies are being used up in making the various products of our throwaway society.

Recycling saves energy. Recycling 1 ton of newspaper saves the amount of energy contained in 100 gallons of gasoline. Recycling 100 pounds of aluminum cans saves an amount of electricity equal to what a typical home uses in six months. Recycling 1 ton of glass saves the equivalent of 10 gallons of gasoline. (This may not seem like a great savings per individual ton. However, when you consider the thousands of tons used *daily* in the United States, it amounts to a significant savings.) When 1 ton of plastic is recycled, it saves the equivalent of 200 to 250 gallons of gasoline.

The amount of energy used by an average American community to cope with its garbage—say, 100 tons of garbage per day for five days per week—is astounding. For example, 43,000 gallons of gasoline are used to run an average landfill per week. In collecting this weekly garbage, 30,000 gallons of gasoline are used, and 13,000 gallons are used in the landfilling process itself.

With voluntary recycling, an average of 255,000 gallons of gasoline are saved weekly. This is in addition to the saving of usable land and the lessening of pollution from the landfill and from manufacturing processes. Mandatory (required by law) recycling saves even more energy than voluntary recycling. It saves an average of 555,000 gallons of gasoline per week. Mandatory composting of yard waste alone not only reduces the need for landfills, it also reduces methane gas pollution and saves 44,000 gallons of gasoline weekly.

By far, the most controversial energy-saving garbage-disposal method is energy recovery through incineration. This method reduces the need for landfill space, saves energy, and produces on average the equivalent of 800,000 gallons of gasoline weekly.

Resource-recovery plants, as some incinerators are called, became very popular after the energy crisis of the 1970s. It was believed that they could offer clean energy—steam—while at the same time reducing trash volume by 90 percent. However, it seems that in spite of sincere efforts to prevent it, most incinerators continue to release toxic pollutants, such as dioxins, into the air. These emissions can cause cancer.

Incinerator operators argue that new pollution controls, such as high-temperature furnaces and so-called *scrubbers* installed in their facilities, eliminate most of the harmful emissions. However, a number of the new incinerators, including one in Detroit, have repeatedly failed air pollution tests, producing too much mercury and toxic ash.

Recycling seems to be a much more practical method for solving today's garbage-energy crisis. It conserves energy, creates new jobs, and protects our natural resources and the environment.

## Science Project #9—Burning Garbage

### *Materials Needed*

- pieces of garbage all about the same size: paper, glass, aluminum, cloth, bread, hard plastic (2-liter soda bottle), apple core
- ring stand with ring
- alcohol burner
- goggles
- wooden kitchen matches
- tin or aluminum dish

(NOTE: THIS EXPERIMENT IS POTENTIALLY DANGEROUS and should be done ONLY with an adult present.)

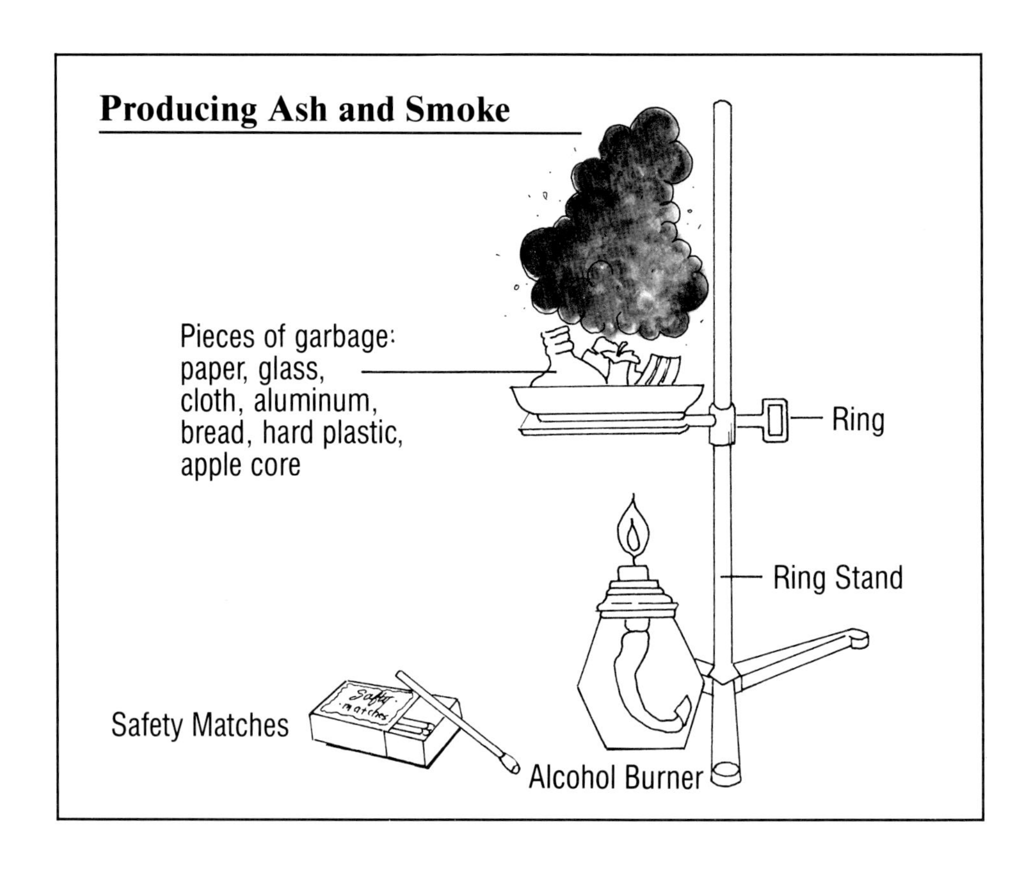


***Prediction.*** Burning garbage will create much smoke and ash.

***Hypothesis.*** If we incinerate garbage, then we will have to deal with the polluting smoke and the ash left behind.

***Procedure.*** Gather samples of garbage, including food scraps, and cut them into small pieces, all about the same size. Set up your ring stand and alcohol burner in a well-ventilated room. Remember to follow all the safety rules used with fire. Tie your hair back and wear goggles.

With an adult present, light your alcohol burner and begin your experiment. Burn each type of garbage in the tin or aluminum dish until nothing is left but ashes. Record your results on a chart.

**Chart of Results**

| | Time to Burn | Descript. of Smoke | Ash | Description after Burning |
|---|---|---|---|---|
| Paper | | | | |
| Glass | | | | |
| Aluminum | | | | |
| Cloth | | | | |
| Bread | | | | |
| Hard Plastic | | | | |
| Apple | | | | |

**(MAKE YOUR OWN CHART OR PHOTOCOPY THIS ONE.)**

## RECYCLING PAPER

Each year, Americans use over 60 million tons of paper. This amounts to an average of 600 pounds per person. The United States is the largest consumer of paper in the world. Paper use from 1965 to the present has doubled.

It takes 17 trees to make 1 ton of paper. That is a lot of trees. It is estimated by the National Association of Recycling Industries that, at current recycling levels, over 200 million trees are spared each year. But each year our need for paper increases. Right now, paper products use up 35 percent of the world's yearly wood harvest. Estimates show that, by the year 2000, the need for paper products will take up 50 percent of the harvest. If we keep cutting our trees for paper, many of the world's forests could be in trouble. Trees cut for paper must be replaced with new trees.

Producing paper from trees also requires a great deal of energy. To produce 1 ton of paper from virgin wood pulp requires 16,320 kilowatt hours (KWH) of electricity. To produce 1 ton of recycled paper from waste paper requires 5,919 KWH. This is a saving of 10,301 KWH (64 percent), a significant amount of energy. The paper industry is the third-largest consumer of energy in the United States. Think about how much energy could be conserved by more recycling of paper.

Recycling not only reduces the amount of energy consumed, it also reduces the amount of air and water pollution that comes from the manufacture of paper from virgin wood. Recycling paper creates at least 35 percent less air pollution than manufacturing paper from virgin wood. It also reduces the amount of solid waste needed to be landfilled, and it uses 58 percent less water.

There are as many different grades of recycled paper as there are papers made from virgin wood. Recycled paper can be of the same quality as paper made from virgin wood pulp. Often, recycled paper has certain characteristics, such as flexibility and density, that make it more desirable than paper made from virgin wood pulp.

*Twenty-five tons of newspaper wait to be shipped out for recycling.*

## Science Project #10—
## Recycling Paper

### *Materials Needed*

- bowl or vat, 24″ by 36″
- distilled water
- newspaper
- scissors
- framed section of screening about 6″ by 6″
- manual or electric blender
- towels or blotting paper
- sponge
- rolling pin
- cheesecloth
- iron

(NOTE: Because of the use of the blender and iron, this experiment is potentially dangerous; an adult should be present to assist.)

***Inference.*** Waste paper can be recycled.

***Hypothesis.*** If waste paper is saved, then it can be made into pulp and recycled.

***Procedure.*** Fill a large bowl or vat with distilled water. Cut up newspapers into thin slivers, and put the paper into the distilled water. Using your blender, beat the paper for four minutes. Add more water and beat again for 90 seconds. Pour the water and chopped paper mixture back into the bowl.

Buy the screen from a hardware store, or cut a 7-inch by 7-inch piece of screening. Frame it with four 1-inch by 2-inch pieces of wood 7 inches long. Use a staple gun or tacks to attach the screen to the wood.

Dip the screened frame into the bowl of water and chopped newspaper and shake it slightly. The paper pulp will cling to the screening. Lift the screening from the water and allow the excess water to drip back into the bowl. Place a towel or blotting paper over the pulp on the screen to absorb the excess water from the pulp. By pressing down gently on the pulp, you will force some of the excess water into the bowl while additional water is absorbed by the towel. Use a sponge to absorb even more of the water, then carefully lift the paper from the screen.

Place the paper down on a dry towel and cover it with another towel. Press down gently in order to remove still more water from the paper. Use a rolling pin to squeeze out even more. Roll the pin back and forth gently over the paper.

Cover the new sheet of paper with wet cheesecloth and place it between two wet blotters. Now iron one side of the blotter "sandwich," turn it over, and iron the other side. Do this for approximately three to four minutes. Remove the paper and trim it to even off the edges.

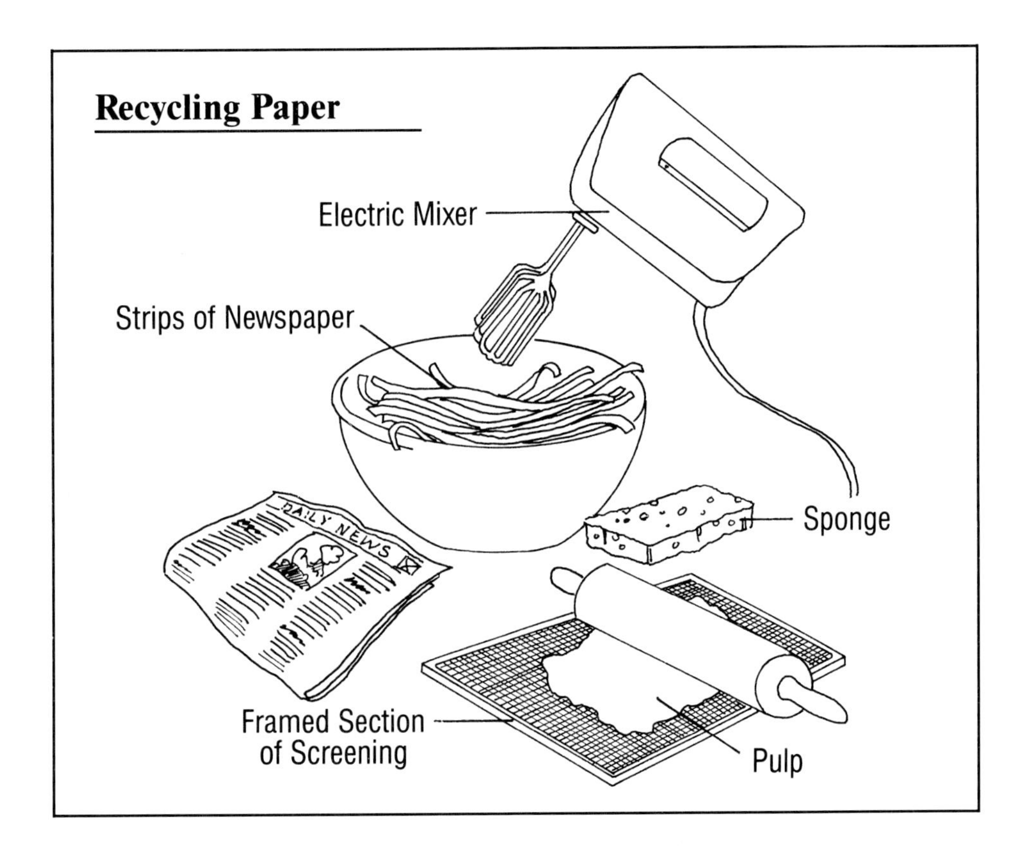

# 6

## *The Future*

With every passing year, we find ourselves buried deeper and deeper in our own garbage. Although we have made some progress in reducing this glut, we have barely begun to scratch the surface.

What exactly is the problem? Is it too much garbage, or is it a thought process that must be changed, along with our behavior? In fact, it is both. One cannot be separated from the other.

Before we can even begin to solve the problems created by our throwaway society, we must change the attitudes that have led to the current situation. For years, citizens of industrialized economies, especially the United States, have believed that they could keep on consuming raw materials without any serious consequences. And it did not seem to matter how much garbage we created, because there was plenty of space to get rid of it.

As we can see, this attitude has led to serious problems. We have too much garbage, we are using up our raw materials, we are polluting our air and water, and we are running out of cheap energy. It is time to realize that we share this Earth with other living things that have an equal right to exist and on which we must depend for our own well-being. Besides being environmentally aware, we must also have an environmental conscience. We have a responsibility, as intelligent creatures, to act for the betterment of the environment, not for its destruction.

How do we go about this? The first way is education. We must become aware of how the environment functions, how nature's cycles work. Once we are educated, we will be better able to appreciate how our actions can affect and alter these cycles.

Second, we must *want* to change our attitudes and behaviors. Knowing how nature functions must be accompanied by sincerely desiring to preserve it for other living things and for future generations of humans.

We must begin now. You can start by going further than the investigations we have done here. Read more about the problem, and see if you can find new ways to solve it. Conserve and recycle. Examine purchases, and buy fewer products whose manufacture is harmful to natural cycles. Try to recover energy without harming the environment.

Write letters to public officials, join environmental organizations, and speak out on environmental issues. If we do these things, we will go a long way toward saving our environment.

# *Conclusions to Projects*

**2.** My hypothesis was correct because the people involved in this experiment did come to better understand the need for waste reduction and recycling.

**3.** My hypothesis was correct because, analyzing the soil samples, I saw a variety of living organisms.

**4.** My hypothesis was correct because the earthworm does utilize the decaying matter in the soil to carry out its activities, which replenishes the soil for plants and other living things.

**5.** My hypothesis was correct because much of the landfilled trash did decompose, while some did not.

**6.** My hypothesis was correct because the seepage from the landfill did contaminate the groundwater.

**7.** My hypothesis was correct because green plants did need carbon dioxide to complete the oxygen cycle.

**8.** My hypothesis was correct because making compost did reduce the amount of garbage to be landfilled.

**9.** My hypothesis was correct because burning garbage did produce a great deal of smoke and ash.

**10.** My hypothesis was correct because waste paper was recycled.

# *For Further Information*

## BOOKS

Earth Works Group. *50 Simple Things Kids Can Do to Save the Earth.* Kansas City, MO: Earthworks Press, 1990.

————————. *50 Simple Things Kids Can Do to Recycle.* San Diego: Greenleaf Classics, Inc., 1991.

Hawkes, Nigel. *Toxic Waste and Recycling.* New York: Gloucester Press, 1988.

Kiefer, Irene. *Poison Land.* New York: Atheneum, 1981.

Lee, Sally. *The Throwaway Society.* New York: Franklin Watts, 1990.

Miller, Christina G., and Louise A. Berry. *Wastes.* New York: Franklin Watts, 1986.

Pringle, Laurence. *Throwing Things Away.* New York: Crowell, 1986.

Weiss, Malcolm. *Toxic Waste.* New York: Franklin Watts, 1984.

Woods, Geraldine and Harold. *Pollution.* Franklin Watts, 1985.

## CONSUMER GROUPS

Environmental Action Foundation
724 Dupont Circle Building NW
Washington, DC 20036
202-745-4871

Environmental Defense Fund
1525 Eighteenth Street NW
Washington, DC 20036
202-387-3500

Greenpeace Action
96 Spring Street
New York, NY 10014
212-941-0994

Natural Resources Defense Council
1350 New York Avenue NW, Suite 300
Washington, DC 20005
202-783-7800

Worldwatch Institute
1776 Massachusetts Avenue NW
Washington, DC 20046
202-452-1999

## GOVERNMENT AGENCIES

Environmental Protection Agency
Public Information Center
401 M Street SW
Washington, DC 20235
202-475-7751

U.S. Department of Energy
1000 Independence Avenue SW
Washington, DC 20585
202-586-5000

U.S. Department of the Interior
1800 D Street NW
Washington, DC 20235
202-208-3100

Individual states' Department of Natural Resources are usually located in the state capital.

# *Glossary*

**Acid.** Any substance that has a sour taste, turns blue litmus paper red, usually contains hydrogen, is corrosive, and measures below 7.0 on the pH scale.

**Bacteria.** One-celled organisms that assist in the decomposition of organic matter.

**Base.** Any substance that has a bitter taste, turns red litmus paper blue, measures 8 to 14 on the pH scale, usually contains a hydroxyl molecule (OH), and is slippery to the touch.

**Biochemical Cycle.** The chemical exchange of mineral elements to organic matter and back.

**Biodegradable.** Wastes that can be broken down in the natural cycle of decomposition.

**Biome.** A single ecological region and its community of living organisms.

**Biosphere.** The portion of the Earth and its atmosphere that can support life.

**Carbon Cycle.** The manner in which carbon is cycled, through the food web, to all living things.

**Classification.** The grouping of objects and information according to their properties.

**Compost Heap.** A busy complex of microbes and other small creatures breaking down organic wastes.

**Conclusion.** The gathering and interpreting of data in a scientific experiment, in order to find out whether a hypothesis is correct or incorrect.

**Decompose.** To break down; organic waste is decomposed mostly by microbes.

**Ecology.** The study of how all living things interrelate with each other and with their nonliving environment.

**Ecosystem.** A specialized community, including all of its various organisms, that functions as an interacting system; one example is a bog.

**Food Web.** Also called the *food chain;* a way of transferring and transforming solar energy for use by a variety of organisms within a community.

**Fossil Fuel.** Any fuel formed from the fossil remains of plants or animals, including coal, oil, and natural gas.

**Fungus.** A plant that contains no chlorophyll and is an important participant in the decomposition of organic waste.

**Groundwater.** Water standing in or moving through the soil and underlying rocks.

**Hypothesis.** An educated guess formed from the observations and classifications concerning a scientific experiment or problem; an inference or prediction that can be tested.

**Incineration.** The process of burning garbage; often used to recover energy from solid waste.

**Inference.** An educated guess about something that has happened based on what has been observed about an event.

**Landfill.** See **Sanitary Landfill.**

**Leachate.** A liquid that dissolves and washes away minerals and perhaps toxic materials out of solids.

**Microbe.** See **Microorganism.**

**Microorganism.** Often called *microbe;* organisms (for example, bacteria, fungi, and protozoans) that are so tiny they can only be seen under a microscope and which aid in the decomposition of organic matter.

**Natural Pollution.** Pollution created by nature through its natural cycles.

**Nitrogen Cycle.** The way in which nitrogen is recycled among living organisms.

**Nonrenewable.** Describes any substance available in only limited quantities; includes substances that take so many years to develop that they are considered nonreplaceable.

**Observation.** A method, using the five senses, of studying scientific or natural phenomena.

**Oxygen Cycle.** The natural cycle by which oxygen, released by plants during photosynthesis, is used by other living things.

**Photosynthesis.** The process by which green plants produce their own food, glucose, from carbon dioxide and sunlight.

**Prediction.** An educated guess, based on observation, about something that is going to happen.

**Protozoan.** A one-celled animal that aids in the decomposition of organic matter.

**Recycling.** Any method of reprocessing and reusing products.

**Renewable.** Describes any resource that can be replaced in a reasonable amount of time.

**Results.** Part of the scientific method; the logging of what happened in an experiment.

**Sanitary Landfill.** A place where solid waste is dumped, compacted into layers, and covered with soil.

**Scrubbers.** Devices put on smokestacks to cleanse emissions.

**Solid Waste.** All solid and semisolid wastes, including trash, garbage, yard wastes, ashes, industrial waste, swill, demolition and construction wastes, and household discards such as appliances, furniture, and equipment.

**Variables.** Differing conditions (such as temperature, time of day, and amount of sunlight) that might affect the outcome of a scientific investigation or experiment.

# *Index*